T0178089

Young Entrepreneurs in Sub-Saharan Africa

Young people in sub-Saharan Africa are growing up in rapidly changing social and economic environments which produce high levels of un- and underemployment. Job creation through entrepreneurship is currently being promoted by international organisations, governments, and NGOs as a key solution, despite there being a dearth of knowledge about youth entrepreneurship in an African context.

This book makes an important contribution by exploring the nature of youth entrepreneurship in Ghana, Uganda, and Zambia. It provides new insights into conceptual and methodological discussions of youth entrepreneurship as well as presenting original empirical data. Drawing on quantitative and qualitative research, conducted under the auspices of a collaborative, interdisciplinary, and comparative research project, it highlights the opportunities and challenges young people face in setting up and running businesses. Divided into a number of clear sections, each with its own introduction and conclusion, the book considers the nature of youth entrepreneurship at the national level, in both urban and rural areas, in specific sectors – including mobile telephony, mining, handicrafts, and tourism – and analyses how key factors, such as microfinance, social capital, and entrepreneurship education, affect youth entrepreneurship.

New light is shed on the multi-faceted nature of youth entrepreneurship and a convincing case is presented for a more nuanced understanding of the term entrepreneurship and the situation faced by many African youth today. This book will be of interest to a wide range of scholars interested in youth entrepreneurship, including in development studies, business studies, and geography, as well as to development practitioners and policy makers.

Katherine V. Gough is Professor of Human Geography at Loughborough University. She has over 20 years' experience of conducting research on urban issues in the global South, with a particular focus on young people.

Thilde Langevang is Associate Professor of Entrepreneurship and Development Studies at Copenhagen Business School. She has been conducting research on youth in Africa for over 10 years.

Routledge spaces of childhood and youth series
Edited by Peter Kraftl and John Horton

The *Routledge Spaces of Childhood and Youth Series* provides a forum for original, interdisciplinary and cutting edge research to explore the lives of children and young people across the social sciences and humanities. Reflecting contemporary interest in spatial processes and metaphors across several disciplines, titles within the series explore a range of ways in which concepts such as space, place, spatiality, geographical scale, movement/mobilities, networks and flows may be deployed in childhood and youth scholarship. This series provides a forum for new theoretical, empirical and methodological perspectives and ground-breaking research that reflects the wealth of research currently being undertaken. Proposals that are cross-disciplinary, comparative and/or use mixed or creative methods are particularly welcomed, as are proposals that offer critical perspectives on the role of spatial theory in understanding children and young people's lives. The series is aimed at upper-level undergraduates, research students and academics, appealing to geographers as well as the broader social sciences, arts and humanities.

Published:
Young Entrepreneurs in Sub-Saharan Africa
Edited by Katherine V. Gough and Thilde Langevang

Forthcoming:
Youth Activism and Solidarity
The non-stop picket against apartheid
Gavin Brown and Helen Yaffe

Children, Young People and Care
Edited by John Horton and Michelle Pyer

Children, Securitization, War and Peace
Kathrin Horschelmann

Children, Nature and Cities
Claire Freeman and Yolanda van Heezik

Young Entrepreneurs in Sub-Saharan Africa

Edited by Katherine V. Gough and Thilde Langevang

Routledge
Taylor & Francis Group

LONDON AND NEW YORK

First published in paperback 2017

First published 2016
by Routledge
2 Park Square, Milton Park, Abingdon, Oxon OX14 4RN

and by Routledge
711 Third Avenue, New York, NY 10017

Routledge is an imprint of the Taylor & Francis Group, an informa
business

British Library Cataloguing in Publication Data
A catalogue record for this book is available from the British Library

Library of Congress Cataloging in Publication Data
Names: Gough, Katherine V., 1963– editor. | Langevang, Thilde, editor.
Title: Young entrepreneurs in Sub-Saharan Africa / edited by Katherine V.
Gough and Thilde Langevang.
Description: Abingdon, Oxon; New York, NY: Routledge, 2016. | Series:
Routledge spaces of childhood and youth series
Identifiers: LCCN 2015035824 | ISBN 9781138844599 (hardback) | ISBN
9781315730257 (e-book)
Subjects: LCSH: Entrepreneurship–Africa, Sub-Saharan. | Young
businesspeople–Africa, Sub-Saharan. | Youth–Employment–Africa,
Sub-Saharan. | Small business–Africa, Sub-Saharan.
Classification: LCC HC800.Y68 2016 | DDC 338/.0408350967–dc23
LC record available at http://lccn.loc.gov/2015035824

ISBN: 978-1-138-84459-9 (hbk)
ISBN: 978-1-138-70493-0 (pbk)
ISBN: 978-1-315-73025-7 (ebk)

Typeset in Times New Roman
by Wearset Ltd, Boldon, Tyne and Wear

In memory of Dr Francis Chigunta, a true scholar and champion of young African entrepreneurs.

Contents

Figures

Tables

Contributors

Robert L. Afutu-Kotey is an urban geographer and Lecturer at the University of Professional Studies, Accra, Ghana (UPSA). He holds an MSc in Regional and Urban Planning from the London School of Economics and Political Science (LSE), and a PhD in Development Studies from the University of Ghana. His doctoral thesis was on *Entrepreneurship and youth livelihoods in the mobile telephony sector in the Greater Accra Metropolitan Area* and was conducted under the YEMP project.

Waswa Balunywa is Professor of Entrepreneurship and the Principal of Makerere University Business School (MUBS), Kampala, Uganda. He holds a PhD in Entrepreneurship from the University of Stirling, UK. His teaching, research, and publication interests are primarily in entrepreneurship, leadership, strategy, and change management, focusing on women, youth, and family business. He has authored books on business administration and business communication skills, and has a chapter in a family business book entitled *Opportunities and dilemmas of social capital: Insights from Uganda*.

Edith Mwebaza Basalirwa is Senior Lecturer of Entrepreneurship and the Acting Dean of the Faculty of Entrepreneurship and Business Administration, Makerere University Business School, Kampala, Uganda. Her primary research interests lie in the field of entrepreneurship, with a particular focus on entrepreneurship education, training, youth, and women. She conducted her PhD studies on *Perceived value of entrepreneurship education, social competences and the entrepreneurial proclivity of university graduates in Uganda* through the YEMP project.

Torben Birch-Thomsen is Associate Professor at the Department of Geosciences and Natural Resource Management, Geography Section, University of Copenhagen, Denmark. He is a human geographer whose key research interests are the environmental and socioeconomic effects of land-use change and intensification. In particular his research focuses on the relations between livelihood strategies in rural communities, farming systems, land-use changes, and rural–urban linkages, especially in East, West, and Southern Africa.

Francis Chigunta is Lecturer in Development Studies and Political Economy at the University of Zambia; he is also Professor of Development Studies at the University of Lusaka and a consultant who regularly works on youth, entrepreneurship, and governance concerns. Between 2009 and 2011 he served first as Chief Policy Analyst, and then as Special Assistant to the President of Zambia (Political). He is the current National Coordinator of the Global Entrepreneurship Monitor (GEM) Project in Zambia, and was the National Coordinator for Zambia on the YEMP project.

Samuel Dawa is a Senior Lecturer at Makerere University Business School, Kampala, Uganda. He is a PhD candidate at the Gordon Institute of Business Science, University of Pretoria, South Africa, where he is doing research on effectuation and sustainable entrepreneurship. His work on youth and women entrepreneurship is published in a range of journals including *International Development Planning Review*, *Journal of Developmental Entrepreneurship*, and *International Journal of Entrepreneurship and Small Business*.

Katherine V. Gough is a Professor of Human Geography at Loughborough University and the principal investigator on the YEMP project, which was based at the University of Copenhagen where she was formerly Associate Professor in the Department of Geography. She has conducted extensive research on urban issues in the global South in a range of contexts in sub-Saharan Africa, Latin America, and the Asia-Pacific region, with a particular focus on comparative and longitudinal studies in low-income settlements. One of her key research interests is young people and she is a co-author of *Youth and the city in the global South: urban lives in Brazil, Vietnam and Zambia.*

Søren Jeppesen is Associate Professor of Small and Medium Enterprises, Corporate Social Responsibility and Entrepreneurship at Copenhagen Business School. He has conducted extensive research on urban and industry related issues in southern and eastern Africa with an emphasis on small firms, corporate social responsibility, and upgrading in selected manufacturing and service industries. He has published his research in a number of journals including *Journal of Business Ethics*, *Business & Society*, *Business Strategy and the Environment*, and *African Journal of Economic and Management Studies*.

Marshall Kala is a Doctoral Fellow at the Institute of Statistical, Social and Economic Research (ISSER), University of Ghana. His main research interests include human development, land tenure, urban development, environmental health, and the extractive sector. His PhD research, conducted as part of the YEMP project, is on the entrepreneurial opportunities for young people in small-scale mining in Ghana. He is a co-author of *Environmental health and disaster risks, livelihoods and ecology within the Korle-Lagoon complex in Accra.*

Søren Bech Pilgaard Kristensen is Associate Professor of Human Geography at the University of Copenhagen. He has conducted extensive research on rural and agriculture-related issues in Uganda with an emphasis on youth and the role of nonfarming activities, and migration from rural to urban areas. He has also conducted research on the consequences and drivers of rural land-use change in Botswana and Mali. His research has been published in a range of journals including *International Development Planning Review* and the *Danish Journal of Geography*.

Sarah Kyejjusa is a Senior Lecturer in the Department of Accounting and a Deputy Director at the Entrepreneurship Centre of Makerere University Business School, Kampala, Uganda. She is a member of the Ugandan Global Entrepreneurship Monitor (GEM) team and an author of their yearly reports. She has conducted research on female entrepreneurship and youth, and has extensive experience of running training courses on business and entrepreneurship for small and medium enterprises (SMEs) in a sub-Saharan African context.

Thilde Langevang is Associate Professor of Entrepreneurship and Development Studies at Copenhagen Business School. She has conducted extensive research on young people and entrepreneurship in a number of African countries. Her research has been published in a range of international journals including *International Development Planning Review*, *Geoforum*, *Children's Geographies*, and *Economic Geography*.

Moonga H. Mumba is Lecturer and Head of the Department of Development Studies at the University of Zambia. Besides teaching, he has been involved in research and consultancy for the Zambian government and local and international development agencies on: youth employment, civil society, governance, land issues, and the administration of justice. He successfully completed his PhD project on *Informal networks and youth self-employment in Zambia* as part of the YEMP project.

Valentine Mwanza is a Lecturer at the Department of Development Studies at the University of Zambia, who has a background in finance, development, and business studies. His key research areas include development finance, investment, youth employment, and entrepreneurship. He is conducting doctoral research on youth employment and entrepreneurship in Zambia. He has participated in the Zambian Global Entrepreneurship Monitor (GEM) team.

Rebecca Namatovu is a Senior Lecturer at Makerere University Business School, Kampala, Uganda and a PhD candidate at the Gordon Institute of Business Science, University of Pretoria, South Africa. She was the National Coordinator for Uganda on the YEMP project. Her entrepreneurship research interests are resource mobilisation, venture growth, youth, and women. Her work has been published in the *Journal of Business Venturing*, *Academy of Management Learning and Education*, *Journal of Developmental Entrepreneurship*, *International Journal of Entrepreneurship and Small Business*, and *International Development Planning Review*.

Agnes Noelin Nassuna is a Lecturer in the Department of Finance and is Deputy Dean in the Faculty of Graduate Studies and Research, Makerere University Business School, Kampala, Uganda. She holds a PhD in Business Administration (Finance) from Makerere University, which was conducted under the auspices of the YEMP project. Her main teaching and research areas are in corporate finance, microfinance, personal, and entrepreneurial finance, with a particular focus on innovative financial approaches to vulnerable groups including low-income groups, women, and young people.

Robert Darko Osei is a Senior Research Fellow at the Institute of Statistical, Social and Economic Research (ISSER), University of Ghana. During 2015 he was Coca-Cola World Fund Visiting Professor of African Studies at the MacMillan Centre for International and Area Studies, Yale University, USA, and a Visiting Scholar at UNU-WIDER, Finland. His main areas of research include evaluative poverty and rural research, macro and micro implications of fiscal policies, aid effectiveness, and other economic development policy concerns.

George Owusu is Associate Professor at the Institute of Statistical, Social and Economic Research (ISSER), and the Head of the Department of Geography and Resource Development, University of Ghana. He was the National Coordinator for Ghana on the YEMP project. His key research interests include urban and regional development, local governance and decentralisation in Ghana and he has published extensively in national and international journals. He is also active on the policy front including participating in developing Ghana's National Urban Policy and Action Plan (2012) as well as the Ghana Urbanization Review (2015), a World Bank/Government of Ghana initiative.

Paul W. K. Yankson is Professor at the Department of Geography and Resource Development, University of Ghana. He is a geographer and an urban planner who has researched on a wide range of urban and regional development issues in Ghana and has published in both local and international journals. He has consulted for both national and international organisations on urban related development issues, including being actively involved in the preparation of Ghana's National Urban Policy and Action Plan (2012) and Ghana Urbanization Review (2015), a World Bank/Government of Ghana initiative.

Acknowledgements

This book is the outcome of the research project "Youth and employment: the role of entrepreneurship in African economies" (YEMP) which ran from October 2009 until June 2014. We are very grateful to the Consultative Research Committee for Development Research (FFU) of the Danish Ministry of Foreign Affairs (Danida) for providing the funding (project number 09-059KU) that made this collaborative and interdisciplinary research possible.

The project involved participants from five institutions located in four countries – Denmark, Ghana, Uganda, and Zambia – all of which provided invaluable support. The project country coordinators, Dr George Owusu at the University of Ghana, Rebecca Namatovu at Makerere University Business School (MUBS), and Dr Francis Chigunta at the University of Zambia, all competently lead their respective teams. In the field we were assisted by numerous research assistants, all of whom did an excellent job of guiding and translating where necessary. This book would not have been possible, though, without all the young people who welcomed us into their homes and workplaces, and generously gave us their time and shared their experiences with us. To all of them, and the adults whom we also interviewed, we owe the greatest thanks.

The project supported six African PhD students who conducted their studies at their home universities but spent six months studying in Copenhagen. Many members of the YEMP team played key roles supervising them, either officially or unofficially, with thanks especially due to Dr Søren Jeppesen at Copenhagen Business School (CBS) and Professor Paul W. K. Yankson, Dr George Owusu and Dr Robert D. Osei at the University of Ghana. In addition, at the University of Zambia, Dr Augustus Kapungwe of the Department of Population Studies supervised the two Zambian PhD students, while at MUBS first the late Dr Warren Byabashaija followed by Professor Waswa Balunywa provided supervision for the two Ugandan PhD students. We are also very grateful to the Danish Fellowship Centre (DFC) for arranging the visits of the PhD students to Denmark and for providing them with an excellent environment in which to live. The Department of Geography, University of Copenhagen, and the Department of Intercultural Communication and Management, CBS, kindly provided the students with office space and ICT support. This conducive environment contributed to the PhD students forming a very supportive group that played a major role in almost all of them succeeding in completing their PhDs, without doubt one of the project's major achievements and a core part of this book.

Many of the chapters in this book were first presented either at our yearly workshops held in the project countries or by YEMP team members at our "International Conference on Entrepreneurship and Employment in the Global South" held in Copenhagen in June 2013. Our excellent keynote speakers, Professor Ernest Aryeetey, Professor Craig Jeffrey, and Professor Peter Rosa, along with all the other participants, provided useful feedback and encouraged us to proceed with presenting the key project findings in book form.

A string of people provided support for the project and the book along the way. At the host institution – the Department of Geography, University of Copenhagen – the untiring assistance of Dorthe Hallin with the accounts was invaluable, and in the closing stages Kent Pørksen kindly helped produce the maps at short notice. Wisdom Kalenga and Cecilia Gregersen both spent time at CBS providing assistance with the data analysis and conference support. Maheen Pracha did an excellent job editing the entire manuscript while Jo Woods assisted in the final checking and layout. Thanks are also due to Faye Leerink at Routledge for seeing the potential of the book and for agreeing to allow it to be subsequently published in sub-Saharan Africa, which we hope will ensure it is also widely read there.

As all of the project participants spent lengthy periods of time in the field and/or visiting other academic institutions, many families have had to cope with these absences. We thank them for their forbearance and for supporting the respective team members in their studies and travels. Hopefully they feel it was worthwhile in the end.

While producing this book has been a major effort, it marks the end of an era that started back in November 2008 when we first started devising the project in response to a call from FFU for projects on youth employment. We are extremely grateful to all of the "YEMP family", as the project team came to be known, for their dedication to the project and for making it such a rewarding and fun experience, and look forward to future collaboration.

Katherine V. Gough and Thilde Langevang
Copenhagen
August 2015

Abbreviations

ALMPs	active labour market policies
AMA	Accra Metropolitan Area
AMFIU	Association of Microfinance Institutions of Uganda
APS	Adult Population Survey
CBOs	community-based organisations
CEEC	Citizens Economic Empowerment Commission
CHOGM	Commonwealth Heads of Government Meeting
CYP	Chawama Youth Project
ELA	Empowerment and Livelihood for Adolescents
FAO	Food and Agriculture Organisation
FDI	foreign direct investment
FGDs	focus group discussions
FISP	Farmer Input Support Programme
GDP	gross domestic product
GEM	Global Entrepreneurship Monitor
GH₵	Ghanaian cedis
GSGDA	Ghana Shared Growth Development Agency
GYEEDA	Ghana Youth Employment and Entrepreneurial Development
ICT	information and communications technology
ILO	International Labour Organization
ISSER	Institute of Statistical, Social and Economic Research
IYF	International Youth Foundation
JHS	junior high school
LCMS	Living Conditions and Monitoring Survey
MAAIF	Ministry of Agriculture, Animal Industry and Fisheries
MDG	Millennium Development Goals
MDI	Microfinance deposit-taking institutions
MFIs	Microfinance institutions
MoYS	Ministry of Youth and Sports
MSMEs	micro, small and medium enterprises
MUBS	Makerere University Business School
NBSSI	National Board for Small Scale Industries
NES	National Expert Survey

NGO	nongovernmental organisation
NPAY	National Programme of Action for Youth
NYDC	National Youth Development Council
NYP	National Youth Policy
PLRF	policy, legal and regulatory framework
PRA	Participatory Rural Appraisal
RYE	Rural Youth Entrepreneurship
SACCOs	saving and credit cooperatives
SAPs	structural adjustment policies
SHS	senior high school
SSM	small-scale mining
TASO	The AIDS Support Organisation
TEA	total early-stage entrepreneurial activity
TEVETA	Technical Education, Vocational and Entrepreneurship Training Authority
TEVST	Technical Education and Vocational Skills Training
TVET	Technical and Vocational Education and Training
UBOS	Uganda Bureau of Statistics
UCU	Uganda Christian University
UGX	Ugandan shilling
UNDP	United Nations Development Programme
UYWS	Uganda Youth Welfare Services
YEMP	Youth and Employment Project
YES	Youth Enterprise Support
ZDA	Zambia Development Agency
ZMK	Zambian Kwacha

1 Introduction

Youth entrepreneurship in sub-Saharan Africa

Katherine V. Gough and Thilde Langevang

Introduction

Young people are highly visible throughout urban and rural areas of sub-Saharan Africa, engaging in a wide range of income-generating activities. Young women operating from open-sided shacks or tabletops can be seen selling a range of goods from fruit and vegetables to cosmetics, and offering services such as plaiting hair and sewing clothes. Young men more commonly deal in manufactured goods, including electronic gadgets, and offer services such as car washing and charging mobile phones, though they are also increasingly entering into former female domains (Overå, 2007). These young people are predominant among the "ordinary" entrepreneurs of the global South (Jeffrey & Dyson, 2013) proving themselves skilful at finding economic niches, managing scarce resources, and seizing profitable opportunities within constrained economic environments.

Economic restructuring and the transformation of labour markets have resulted in limited employment opportunities for young people whose unemployment rates can be two to three times higher than the norm (World Bank, 2012). Young people are often depicted as having become increasingly marginalised, causing idleness and frustration which, it is believed, can lead to involvement in crime, organised violence, and protests (Garcia & Fares, 2008; World Bank, 2006, 2012).[1] Consequently, entrepreneurship is increasingly being promoted as a key tool to combat the youth unemployment crisis and as one of the main drivers of economic and social transformation in sub-Saharan Africa (Africa Commission, 2009; World Bank, 2006). In light of their limited possibilities to gain formal sector jobs in the public or private sector, young people are being encouraged to be "job creators" rather than "job seekers", thus becoming self-employed "entrepreneurs" (Langevang & Gough, 2012).

Despite the increasing focus on youth entrepreneurship, little is known about who the young entrepreneurs are, which activities they engage in, the challenges they face, and their aspirations. This book aims to fill this gap by presenting detailed studies of the experiences of young entrepreneurs in a range of settings in sub-Saharan Africa. As well as presenting original empirical data, it also contributes to discussions regarding the concept of entrepreneurship and methodological approaches to studying youth entrepreneurship. The book is the principal outcome of a four-year research project titled "Youth and Employment: The Role of Entrepreneurship in African Economies" (YEMP), which explored youth entrepreneurship in Ghana, Uganda, and Zambia. An interdisciplinary team of 20 researchers from geography, business studies, and development studies worked collaboratively, conducting both quantitative and qualitative research. In this book, we analyse the nature of youth entrepreneurship at the national level in both urban and rural areas and in specific sectors, and show how it is affected by key factors including microfinance, social capital, and entrepreneurship education. We illustrate the multifaceted nature of youth entrepreneurship and argue for a more nuanced understanding of the term "entrepreneurship" and the situation faced by many African youth today.

Conceptualising and linking youth and entrepreneurship

Youth and entrepreneurship are both highly debated concepts. The most common way of defining youth is by chronological age, such as the United Nations' categorisation of youth as 15–24-year-olds. Many sub-Saharan African countries, however, define youth in more expansive terms, pushing the ceiling into the mid-30s (Chigunta, Schnurr, James-Wilson, & Torres, 2005). Consequently, in this project, we conducted research with young people aged between 15 and 35 years. Youth is clearly more than an age bracket, however, and there are numerous ways of conceptualising the term. Two key perspectives are the life-stage – otherwise known as the youth transitions – perspective, and the youth culture perspective (Christiansen, Utas, & Vigh, 2006; MacDonald et al., 2001).

Life-stage or youth transition studies characterise youth as a distinct stage between childhood (a phase of dependence and immaturity) and adulthood (a phase of independence and maturity). Youth is depicted as a stage through which people pass in order to become mature, independent adults (Skelton, 2002). Transition studies, however, have been criticised for failing to comprehend young people's complex life paths in times of rapid socioeconomic change, and for reducing youth to a transitory state of becoming "rather than a recognized stage in its own right with distinctive experiences and issues" (Skelton, 2002: 103). The transitions that have traditionally been associated with growing up, such as finding a job, leaving home, getting married, and becoming a parent, may in fact occur simultaneously or not at all, and many of them are reversible (Johnson-Hanks, 2002; Valentine, 2003). This is especially true in sub-Saharan Africa where rather than passing through neat, linear life-stages, young people traverse back and forth between fluid boundaries of time- and place-specific notions of childhood, youth, and adulthood (Burgess, 2005; Christiansen et al., 2006; Honwana & De Boeck, 2005; Johnson-Hanks, 2002; Langevang, 2008; van Blerk, 2008).

In contrast, cultural approaches to youth focus on young people's here-and-now lived experiences, with a particular interest in their sub-cultures and styles. This perspective sees young people as a socially and culturally demarcated group with distinctive views and experiences. While recognising the importance of understanding how youth experiences vary across different cultures, the cultural approach has been criticised for focusing on deviant, spectacular, and male youth, and for analysing young people's agency in isolation from the surrounding society (Valentine, Skelton, & Chambers, 1998). These studies of youth cultures have been dominated by research in northern contexts, especially Britain and the United States, with few studies conducted in sub-Saharan Africa.

The life-stage/transitions approach and the cultural approach, however, are not mutually exclusive and the two can be usefully combined. As Christiansen et al. (2006) argue, understanding the lives of young people is a question of combining an analysis of how young people see and interpret the world, and how they are positioned within it through their families and societies. This requires focusing simultaneously on young people's agency and how they are restricted by their environments to show how "youth are able to move, what they seek to move towards and the ways external forces seek to shape their movements" (Christiansen et al., 2006: 16). This is the approach this book adopts.

During the past decade, there has been an upsurge of research highlighting the diversity and multiplicity of young people's trajectories and lived experiences in a range of settings in sub-Saharan Africa (see, among others, Ansell, van Blerk, Hajdu, & Robson, 2011; Burgess, 2005; Camfield, 2011; Christiansen et al., 2006; Darkwah, 2013; Day & Evans, 2015; Esson, 2013; Gough, 2008; Hajdu, Ansell, Robson, & van Blerk, 2013; Honwana, 2012; Honwana & De Boeck, 2005; Kristensen & Birch-Thomsen, 2013; Langevang, 2008; Mains, 2012; Porter et al., 2010; Ralph, 2008; Sommers, 2012; Thorsen, 2013; van Blerk, 2008; Weiss, 2009). As these studies highlight, young people's lived experiences vary greatly according to gender, socioeconomic status, education, and place. In this book, we add to this literature by examining how youth entrepreneurship varies by social differences, highlighting both similarities and differences between youth in a range of settings.

Turning to entrepreneurship, the word stems from the French word *entreprendre*, which means "to begin something" or "to undertake". The role of the entrepreneur was first recognised in the 18th century by the economist Richard Cantillon, whose research provided the basis for three major economic traditions (Hébert & Link, 1989). The first was the German tradition dominated by the research of Schumpeter (1934/1983) who saw entrepreneurship as key to economic development and entrepreneurs as extraordinary individuals who introduced new products or processes, identified new markets or sources of supply, or developed new types of organisations. The second, the Chicago tradition led by Knight (1921/2012), saw entrepreneurs as risk takers and argued that opportunity for profit arose because of the uncertainty surrounding change. The third, the Austrian tradition, was proposed by Kirzner (1978), who considered the entrepreneur somebody who was alert to profitable opportunities for exchange that occurred due to information gaps in the market, which they were able to identify because they had superior knowledge.

To date, most entrepreneurship research has been dominated by an economic discourse and focused on elite businesses in the global North (Gough, Langevang,

& Owusu, 2013). More recently, however, in line with Steyaert and Katz's (2004) notion of entrepreneurship as an "everyday societal phenomenon", there has been a widening of the domains, spaces, and discourses of entrepreneurship. Consequently, entrepreneurship is seen as an activity not restricted to a select few individuals but one in which "ordinary" individuals engage. This is the approach adopted in this book where entrepreneurship is seen as spotting and seizing an opportunity to establish a business and, in the process, taking risks and managing the available resources creatively.

We focus in particular on the process-oriented character of entrepreneurship – "entrepreneuring" – which is seen as a practice that affects and is affected by the economic, social, institutional, and cultural environment (Steyaert, 2007), highlighting how different actors, forms, practices, and discourses of entrepreneurship emerge in differing societal contexts. As Baumol (1990: 898) notes, the rules for entrepreneurship "change dramatically from one time and place to another" and context is vital for understanding "when, how, and why entrepreneurship happens and who becomes involved" (Welter, 2011: 166). Hence, although conventional theories of entrepreneurship have tended to focus on the isolated individual entrepreneur, there is now growing acknowledgement that entrepreneurial behaviour must be understood within its historical, temporal, spatial, institutional, and social context (Welter, 2011).

While research on entrepreneurship has been dominated by experiences from the global North, in recent years there has been an upsurge in research on entrepreneurship in the global South – including Africa – spearheaded by Spring and McDade's (1998) pioneering book, *African Entrepreneurship: Theory and Reality*. This research has focused on a broad range of topics, such as entrepreneurs' motivation for starting a business (Rosa, Kodithuwakku, & Balunywa, 2006), the characteristics of family businesses (Khavul, Bruton, & Wood, 2009), the role of social networks and associations (Kuada, 2009; McDade & Spring, 2005), innovation in small enterprises (Robson, Haugh, & Obeng, 2009), and the particular challenges facing female entrepreneurs (Amine & Staub, 2009).

A related field of literature concerns the "informal sector" where most entrepreneurial activity in Africa is located (Lindell, 2010; Potts, 2008). The terms "informal sector", "informal work", and "informal business" are commonly used but difficult to define. The informal sector is typically considered to comprise "economic activities that lie beyond or circumvent state regulation" (Lindell, 2010: 5), though, as several of the chapters in this book illustrate, this is not unproblematic as many informal businesses do pay licence fees and taxes to the local government. Delimiting the boundaries of employment and unemployment is also problematic since unemployment is challenging to measure, track over time, and compare across contexts (Izzi, 2013). This is not least the case in economies characterised by informality where work is often irregular, insecure, and casual.

While there is a growing body of literature on both youth and entrepreneurship in Africa, there are still very few studies of youth entrepreneurship in Africa. Consequently, there is limited knowledge of the level and characteristics of youth entrepreneurial activity, including the types of business into which young people venture and the resources on which they draw (for exceptions, see Langevang & Gough, 2012; Langevang, Namatovu, & Dawa, 2012; Mwasal- wiba, Dahles, & Wakkee, 2012). Despite this lack of knowledge, the notion that youth should become entrepreneurs is being promoted by development organisa- tions and institutions such as the World Bank (2006; see also Africa Commis- sion, 2009) and has been endorsed by governments and institutions throughout sub-Saharan Africa (see Oppenheimer et al., 2011). Consequently, there has been a mushrooming of initiatives, programmes, and policies to foster entrepre- neurship among young people.

Discourses of entrepreneurship have been underpinned by neoliberal market values of competition, efficiency, self-reliance, and self-governance (Dolan, 2012). As Jeffrey and Dyson (2013: R1) highlight, however, there are "ideo- logical risks attached to celebrations of youth entrepreneurialism". The idea that people can "pull themselves up by their bootstraps" by engaging in entrepreneur- ship can result in a lack of state investment in services and in structural problems being seen as the individual's responsibility, resulting in young people blaming themselves for their predicament. Moreover, it is important to recognise that not everyone has the assets – whether financial, social, or bodily – to become entre- preneurs (Dolan, 2012). There is a need, therefore, to understand the entrepre- neurial efforts young people are making, the challenges they face, and to devise appropriate support measures; this book aims to do just this.

Researching youth entrepreneurship across countries

Research on entrepreneurship has been dominated largely by quantitative meth- odologies, which use structured surveys. As Neergaard and Ulhøi (2007: 1) argue, however, the complex and dynamic nature of entrepreneurship requires a "methodological toolbox of broad variety". Using mixed methods is even more relevant in a sub-Saharan African context where much entrepreneurship takes place outside of formal registered businesses (Gough, Langevang, & Namatovu, 2014). By contrast, research on youth is characterised by the widespread adop- tion of qualitative methods, many of which are participatory. A wide range of methods, including the use of drawings, diaries, plays, photographs, etc., as well as the more standard in-depth interviews and focus group discussions are com- monly used when researching young people (Langevang, 2007). Researching "youth entrepreneurship" thus entails bringing together two, often very different, methodological approaches. Inspired by both research on entrepreneurship and youth, in this project we have combined a quantitative approach using structured surveys with a range of qualitative methodologies, including in-depth interviews, focus group discussions, and participatory methods.

Ghana, Uganda, and Zambia were selected for this study of youth entrepre- neurship as they are anglophone African countries located in West, East, and Southern Africa, respectively. Although all three are characterised as "develop- ing" countries, according to the World Bank's classification Ghana and Zambia have now joined the ranks of lower middle-income countries while Uganda remains in the category of low-income countries. Researching in three countries with their respective differing economies and societies provided the possibility to create a greater understanding of the opportunities and challenges facing young entrepreneurs in sub-Saharan Africa today. Within each country, a team of local researchers, including two PhD students, collaborated with Danish researchers at all stages of the research process. The research team was interdisciplinary, consisting of geographers based at the University of Ghana, business studies experts from Makerere University Business School, Uganda, development studies experts based at the University of Zambia, and in Denmark geographers from the University of Copenhagen and business studies experts based at the Copenhagen Business School (see Figure 1.1).

In order to generate a picture of overall levels of youth entrepreneurship as well as an in-depth understanding of the influencing factors, the project was designed to incorporate both quantitative and qualitative data collection. An important aspect of the quantitative data collection was participating in the Global Entrepreneurship Monitor (GEM) survey. GEM is the largest survey of entrepreneurial activity in the world; in 2012, when the GEM studies presented in this book were conducted, a total of 69 countries participated, all of which followed a common methodology (see Introduction to Part I for more details). The GEM survey generated quantitative data on levels of entrepreneurship at the national and regional levels, including detailed information on young entrepreneurs; this data forms the basis for the chapters in Part I of this

Figure 1.1 The YEMP team at the project workshop in Kampala (photo: Torben Birch-Thomsen).

book. In order to generate a more in-depth understanding of youth entrepreneurship, a series of studies were undertaken by the YEMP team. Within each country, urban and rural areas were selected for detailed case studies along with certain sectors of the economy. Most of these studies entailed collecting both quantitative and qualitative data, some of which used participatory methodologies (see the individual chapters for more detail), with common methodologies being adopted across countries where appropriate to facilitate comparison.

When conducting comparative research, it is vital to be aware of what is being compared. In the words of Melhuus (2002: 82), "Because similarities or differences are not given in the things themselves but in the ways they are contextualized", we must "compare meanings, ways of constructing relationships between objects, persons, situations, events". A comparative approach was central to the YEMP research project at a range of scales: between countries at national level and across both rural and urban settings; within countries between rural and urban areas and across sectors; and between youth of differing genders, income levels, and education. Correspondingly, within this book comparisons are drawn at national, settlement, and sectoral levels and the differences between youth with differing characteristics are illustrated. By comparing the similarities and differences between youth experiences of entrepreneurship, we highlight commonalities as well as indicating diversity.

Structure and contributions of book

The book is structured into five separate but interlinked parts, each preceded by an introduction and rounded off with a comparative discussion of the findings. The first part examines youth entrepreneurship at the national level in Uganda, Ghana, and Zambia. Each of the three chapters draws on GEM data to provide a picture of the level and types of youth entrepreneurship in the respective countries. The policies that have been introduced to support youth entrepreneurship are also presented. In the concluding comments to the section, the similarities and differences in youth entrepreneurship between the three countries are drawn out. This section thus sets the scene for the chapters that follow.

The second part explores the nature of youth entrepreneurship in low-income urban settlements. In each country, one low-income settlement located in the capital city was studied, using a similar methodology that entailed working with young researchers to collect quantitative and qualitative data. Each of the three chapters presents and analyses data on the number and types of businesses run by young people in the respective settlements. In the concluding comments, the similarities and differences between youth entrepreneurship in Accra, Kampala, and Lusaka are highlighted.

The third part focuses on youth in rural areas. In each of the project countries, a study was conducted on youth entrepreneurship in one rural area, using a similar methodology. In the three chapters in this section, the nature of youth entrepreneurship in agricultural and nonfarm enterprises is discussed and the reasons these young people remain in, migrate from, or have returned to rural areas is explored. In the concluding comments, these findings are drawn upon to highlight the similarities and differences between youth entrepreneurship in the rural areas studied.

As part of the YEMP project, a number of studies of youth running businesses in specific sectors were conducted. These are presented in the fourth part with individual chapters focusing on four specific sectors, namely young people working in the mobile telephony sector in Accra, Ghana, engaged in small-scale mining in northern Ghana, making and selling handicrafts in western Uganda, and working in the tourism industry in Uganda. In the concluding comments, the challenges and opportunities that young people face in their various business types are highlighted.

The fifth and final part presents three studies that examine ways of stimulating entrepreneurship undertaken under the auspices of the project. First, the role of social capital drawn on by young entrepreneurs in setting up and running businesses in Lusaka, Zambia, is examined. Second, the ways in which microfinance institutions cater to youth in Uganda is presented. Third, the role of education in promoting entrepreneurship is discussed, drawing on the experience of university graduates of entrepreneurship in Uganda. A discussion of the emerging issues for young entrepreneurs in sub-Saharan Africa concludes this part and the book.

Note

1 A causal link between unemployment and violence among youth has not, however, been empirically demonstrated (Izzi, 2013).

References

Africa Commission. (2009). *Realising the potential of Africa's youth: Report of the Africa Commission.* Copenhagen: Ministry of Foreign Affairs of Denmark.

Amine, L. S., & Staub, K. M. (2009). Women entrepreneurs in sub-Saharan Africa: An institutional theory analysis from a social marketing point of view. *Entrepreneurship and Regional Development, 21*(2), 183–211.

Ansell, N., van Blerk, L., Hajdu, F., & Robson, E. (2011). Space, times, and critical moments: A relational time-space analysis of the impacts of AIDS on rural youth in Malawi and Lesotho. *Environment and Planning A, 43*(3), 525–544.

Baumol, W. J. (1990). Entrepreneurship: Productive, unproductive and destructive. *Journal of Political Economy, 98*(5), 893–921.

Burgess, T. (2005). Introduction to youth and citizenship in East Africa. *Africa Today, 51*(3), 7–24.

Camfield, L. (2011). "From school to adulthood?" Young people's pathways through schooling in urban Ethiopia. *European Journal of Development Research, 23*(5), 679–694.

Chigunta, F., Schnurr, J., James-Wilson, D., & Torres, V. (2005). *Being "real" about youth entrepreneurship in Eastern and Southern Africa* (SEED Working Paper No. 72). Geneva: International Labour Office.

Christiansen, C., Utas, M., & Vigh, H. E. (2006). Introduction: Navigating youth, generating adulthood. In C. Christiansen, M. Utas, & H. E. Vigh (Eds), *Navigating youth,
generating adulthood: Social becoming in an African context* (pp. 9–28). Uppsala:
Nordic Africa Institute.

Darkwah, A. K. (2013). Keeping hope alive: An analysis of training opportunities for
Ghanaian youth in the emerging oil and gas industry. *International Development Planning Review, 35*(2), 119–134.

Day, C. & Evans, R. (2015). Caring responsibilities, change and transitions in young people's family lives in Zambia. *Journal of Comparative Family Studies, 46*(1), 137–152.

Dolan, C. (2012). The new face of development: The "bottom of the pyramid" entrepreneurs. *Anthropology Today, 28*(4), 3–7.

Esson, J. (2013). A body and a dream at a vital conjuncture: Ghanaian youth, uncertainty
and the allure of football. *Geoforum, 47*, 84–92.

Garcia, M., & Fares, J. (Eds). (2008). *Youth in Africa's labour market.* Washington, DC:
World Bank.

Gough, K. V. (2008). "Moving around": The social and spatial mobility of youth in
Lusaka. *Geografiska Annaler, 90*(3), 243–255.

Gough, K. V. (2010). Continuity and adaptability of home-based enterprises: A longitudinal study from Accra, Ghana. *International Development Planning Review, 32*(1),
45–70.

Gough, K. V., Langevang, T., & Namatovu, R. (2014). Researching entrepreneurship in
low-income settlements: The strengths and challenges of participatory methods.
Environment and Urbanization, 26(1), 297–311.

Gough, K. V., Langevang, T., & Owusu, G. (2013). Youth employment in a globalising
world. *International Development Planning Review, 35*(2), 91–102.

Hajdu, F., Ansell, N., Robson, E., & van Blerk, L. (2013). Rural young people's opportunities for employment and entrepreneurship in globalised southern Africa: The limitations of targeting policies. *International Development Planning Review, 35*(2),
155–174.

Hébert, R. F., & Link, A. N. (1989). In search of the meaning of entrepreneurship. *Small
Business Economics, 1*(1), 39–49.

Honwana, A. (2012). *The time of youth: Work, social change and politics in Africa.*
Boulder, CO: Kumarian Press.

Honwana, A., & De Boeck, F. (Eds). (2005). *Makers and breakers: Children and youth
in postcolonial Africa.* Oxford: James Currey.

Izzi, V. (2013). Just keeping them busy? Youth employment projects as a peacebuilding
tool. *International Development Planning Review, 35*(2), 103–117.

Jeffrey, C., & Dyson, J. (2013). Zigzag capitalism: Youth entrepreneurship in the contemporary global South. *Geoforum, 49*, R1–R3.

Johnson-Hanks, J. (2002). On the limits of life stages in ethnography: Toward a theory of
vital conjunctures. *American Anthropologist, 104*(3), 865–880.

Jones, J. L. (2010). "Nothing is straight in Zimbabwe": The rise of the Kukiya-kiya
economy 2000–2008. *Journal of Southern African Studies, 36*(2), 285–299.

Khavul, S., Bruton, G. D., & Wood, E. (2009). Informal family business in Africa. *Entrepreneurship Theory and Practice, 33*(6), 1217–1236.

Kirzner, I. M. (1978). *Competition and entrepreneurship.* Chicago, IL: University of
Chicago Press.

Knight, F. H. (2012). *Risk, uncertainty and profit.* Mineola, NY: Dover Publications.
(Original work published 1921)

Kristensen, S., & Birch-Thomsen, T. (2013). Should I stay or should I go? Rural youth employment in Uganda and Zambia. *International Development Planning Review, 35*(2), 175–202.

Kuada, J. (2009). Gender, social networks, and entrepreneurship in Ghana. *Journal of African Business, 10*, 85–103.

Langevang, T. (2007). Movements in time and space: Using multiple methods in research with young people in Accra, Ghana. *Children's Geographies, 5*(3), 267–281.

Langevang, T. (2008). "We are managing!" Uncertain paths to respectable adulthoods in Accra, Ghana. *Geoforum, 39*(6), 2039–2047.

Langevang, T., & Gough, K. V. (2012). Diverging pathways: Young female employment and entrepreneurship in sub-Saharan Africa. *The Geographical Journal, 178*(3), 242–252.

Langevang, T., Namatovu, R., & Dawa, S. (2012). Beyond necessity and opportunity entrepreneurship: Motivations and aspirations of young entrepreneurs in Uganda. *International Development Planning Review, 34*(4), 242–252.

Lindell, I. (2010). Introduction: The changing politics of informality – Collective organizing, alliances and scales of engagement. In I. Lindell (Ed.), *Africa's informal workers: Collective agency, alliances and transnational organizing in urban Africa* (pp. 1–30). London: Zed Books and Nordiska Afrikainstitutet.

MacDonald, R., Mason, P., Shildrick, T., Webster, C., Johnston, L., & Ridley, L. (2001). Snakes and ladders: In defence of studies of youth transition. *Sociological Research Online, 5*(4). Retrieved from www.socresonline.org.uk/5/4/macdonald.html.

Mains, D. (2012). *Hope is cut: Youth unemployment and the future in urban Ethiopia.* Philadelphia: Temple University Press.

McDade, B. E., & Spring, A. (2005). The "new generation of African entrepreneurs": Networking to change the climate for business and private sector-led development. *Entrepreneurship and Regional Development, 17*, 17–42.

Melhuus, M. (2002). Issues of relevance: Anthropology and the challenges of cross-cultural comparison. In A. Gingrich & R. G. Fox (Eds), *Anthropology, by comparison* (pp. 70–91). London and New York: Routledge.

Mwasalwiba, E., Dahles, H., & Wakkee, I. (2012). Graduate entrepreneurship in Tanzania: Contextual enablers and hindrances. *European Journal of Scientific Research, 76*(3), 386–402.

Neergaard, H., & Ulhøi, J. P. (2007). Introduction: Methodological variety in entrepreneurship research. In H. Neergaard & J. P. Ulhøi (Eds), *Handbook of qualitative research methods in entrepreneurship* (pp. 1–14). Cheltenham: Edward Elgar.

Oppenheimer, J., Spicer, M., Trejos, A., Zille, P., Benjamin, J., Cavallo, D., Kacou, E., & Leo, B. (2011). *Putting young Africans to work: Addressing Africa's youth unemployment crisis* (Discussion Paper No. 2011/08). Johannesburg: Brenthurst Foundation.

Overå, R. (2007). When men do women's work: Structural adjustment, unemployment and changing gender relations in the informal economy of Accra, Ghana. *Journal of Modern African Studies, 45*(4), 539–563.

Porter, G., Hampshire, K., Abane, A., Robson, E., Munthali, A., Mashiri, M., & Tanle, A. (2010). Moving young lives: Mobility, immobility and inter-generational tensions in urban Africa. *Geoforum, 41*(5), 796–804.

Potts, D. (2008). The urban informal sector in sub-Saharan Africa: From bad to good (and back again?). *Development Southern Africa, 25*(2), 151–167.

Ralph, M. (2008). Killing time. *Social Text, 26*(4), 1–29.

Robson, P., Haugh, H., & Obeng, B. (2009). Entrepreneurship and innovation in Ghana: Enterprising Africa. *Small Business Economics, 32*(3), 331–350.

Rosa, P. J., Kodithuwakku, S., & Balunywa, W. (2006). Entrepreneurial motivation in developing countries: What does "necessity" and "opportunity" entrepreneurship really mean? *Frontiers of Entrepreneurship Research, 26*, 1–14.

Schumpeter, J. A. (1983). *The theory of economic development.* New Brunswick, NJ: Transaction Publishers. (Original work published 1934)

Skelton, T. (2002). Research on youth transitions: Some critical interventions. In M. Cieslik & G. Pollock (Eds), *Young people in risk society: The restructuring of youth identities and transitions in late modernity* (pp. 100–116). Aldershot: Ashgate.

Sommers, M. (2012). *Stuck: Rwandan youth and the struggle for adulthood.* Athens and London: University of Georgia Press.

Spring, A., & McDade, B. E. (Eds). (1998). *African entrepreneurship: Theory and reality.* Gainesville, FL: University Press of Florida.

Steyaert, C. (2007). "Entrepreneuring" as a conceptual attractor? A review of process theories in 20 years of entrepreneurship studies. *Entrepreneurship and Regional Development, 19*(6), 453–477.

Steyaert, C., & Katz, J. (2004). Reclaiming the space of entrepreneurship in society: Geographical, discursive and social dimensions. *Entrepreneurship and Regional Development, 16*(3), 179–196.

Thorsen, D. (2013). Weaving in and out of employment and self-employment: Young rural migrants in the informal economy of Ouagadougou. *International Development Planning Review, 35*(2), 203–218.

Valentine, G. (2003). Boundary crossings: Transitions from childhood to adulthood. *Children's Geographies, 1*(1), 37–52.

Valentine, G., Skelton, T., & Chambers, D. (1998). Cool places: An introduction to youth and youth cultures. In T. Skelton & G. Valentine (Eds), *Cool places: Geographies of youth cultures* (pp. 1–34). London: Routledge.

van Blerk, L. (2008). Poverty, migration and sex work: Youth transitions in Ethiopia. *Area, 40*(2), 245–253.

Weiss, B. (2009). *Street dreams and hip hop barbershops: Global fantasy in urban Tanzania.* Bloomington and Indianapolis: Indiana University Press.

Welter, F. (2011). Contextualising entrepreneurship: Conceptual challenges and ways forward. *Entrepreneurship Theory and Practice, 35*(1), 165–184.

World Bank. (2006). *World development report 2007: Development and the next generation.* Washington, DC: World Bank.

World Bank. (2012). *World development report 2013: Jobs.* Washington, DC: World Bank.

Part I

National studies of youth entrepreneurship

Introduction to Part I

Katherine V. Gough and Thilde Langevang

Entrepreneurship is widely seen as a driver of economic growth as entrepreneurs create new businesses, introduce innovation, contribute to structural changes in the economy, and introduce new competition through launching new products and services (Kew, Herrington, Litovsky, & Gale, 2013). This vital role has heightened the interest in measuring and comparing levels and characteristics of entrepreneurship around the world as well as in understanding how entrepreneurship levels are influenced by national political, socioeconomic, and institutional environments. Since entrepreneurship is also increasingly seen as a solution to the youth unemployment challenge that many countries are facing, measuring the levels, attitudes, and characteristics of young entrepreneurs as well as understanding the national socioeconomic and policy environment young entrepreneurs are embedded in are pivotal. This part of the book provides an overall picture of the scale and characteristics of youth entrepreneurship at the national level in Ghana, Uganda, and Zambia before presenting key policies and programmes adopted to promote entrepreneurship in the three countries.

The analysis of the scale and characteristics of youth entrepreneurship draws predominantly on quantitative data generated as part of the Global Entrepreneurship Monitor (GEM) survey which defines entrepreneurship as "any attempt at new business or new venture creation, such as self-employment, a new business organisation, or the expansion of an existing business, by an individual, a team of individuals, or an established business" (Kew et al., 2013: 9). GEM was designed to do the challenging task of measuring and comparing entrepreneurship at the national level, and has become the largest, and most highly respected, survey of entrepreneurship in the world. From including just 10 developed economies at its inception in 1999, GEM has grown over the years to cover 69 participating countries in 2012 including a number of sub-Saharan African countries; apart from Ghana, Uganda, and Zambia, the other sub-Saharan participants included Angola, Botswana, Ethiopia, Malawi, Namibia, Nigeria, and South Africa.

GEM adopts a common methodology for all participating countries consisting of an Adult Population Survey (APS) and a National Expert Survey (NES). These chapters predominantly draw on data from the APS, which is an annual survey of entrepreneurship behaviour and attitudes towards entrepreneurship among the population. To ensure consistency and cross-country comparability, each participating country conducts the same questionnaire survey of its adult population (18–64 years old) with a minimum of 2,000 randomly selected individuals. The questionnaire is structured into a number of sections based on a distinction between three types of entrepreneurs: "nascent entrepreneurs" (those who have been paying salaries for less than three months), "new business owner-managers" (those who have been paying salaries for between three and 42 months), and "established business owners" (those who have paid salaries and wages for more than 42 months). A key measurement is the total early-stage entrepreneurial activity (TEA), which includes "nascent entrepreneurs" and "new business owner-managers" (see Figure PI.1 and Xavier, Kelley, Kew, Herrington, & Vorderwülbecke, 2013). The motivation for starting a business, the type of business activity, the degree of innovativeness, size of enterprises in terms of employees, and job growth expectations of the entrepreneurs are all recorded. In addition, non-business owners are asked about their entrepreneurial intentions and all respondents are asked about their attitudes towards entrepreneurship. Basic demographic characteristics of the respondents (education, age, household income, employment status) are also recorded. The chapters in this part predominantly use data for the youth (aged 18–35), who are compared to adults when relevant. While acknowledging that the GEM data is not without limitations especially in a sub-Saharan African context (see Gough, Langevang, & Namatovu, 2014; Rosa, Kodithuwakku, & Balunywa, 2006), we nevertheless find it useful for illuminating overall patterns of youth entrepreneurship in Ghana, Uganda, and Zambia and for comparing the patterns between the countries and with other countries.

The analysis of entrepreneurship policies and programmes builds on mapping exercises carried out in each of the three countries. As Schoof (2006: 67) illustrates, youth entrepreneurship policy can be defined as policy measures taken to

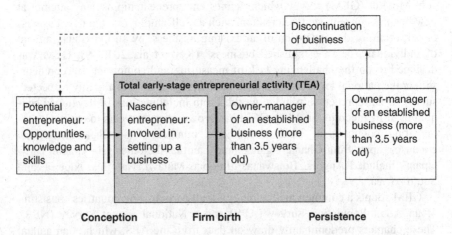

Figure PI.1 GEM operational definitions (source: Kelley et al., 2011).

foster entrepreneurial activity of young people in three differing ways: first, those aimed at the pre-start-up, start-up, and post-start-up phases of the entrepreneurial process; second, policies designed and delivered to address the areas of motivation, opportunity, and skills; and third, those with the main objective of encouraging more people to start and grow an entrepreneurial venture and at the same time improve young people's general employability. In practice, however, it is difficult to outline the entire range of policies and programmes that affect youth entrepreneurship as initiatives often straddle employment policies, educational policies, and private sector and enterprise promotion programmes. Whereas some policies and programmes only target young people, in other programmes young people are treated as part of the general population. Furthermore, entrepreneurship development embraces a range of actors and institutions including educational, regulatory, financial, and business support institutions, and involves the public sector, NGOs, the private sector, and multilateral organisations.

The three chapters in this part of the book provide an overview of key characteristics of each country before examining in depth the entrepreneurship activities of the youth. This is followed by an overview of key youth employment and entrepreneurship policies and programmes, focusing especially on government policies, highlighting their aims and discussing their impact and challenges.

References

Gough, K. V., Langevang, T., & Namatovu, R. (2014). Researching entrepreneurship in low-income settlements: The strengths and challenges of participatory methods. *Environment and Urbanization, 26*(1), 297–311.

Kelley, D. J., Bosma, N., & Amoros, J. E. (2011). *Global Entrepreneurship Monitor (GEM) 2010 Global Report*. GERA/GEM. Retrieved from www.gemconsortium.org/report.

Kew, J., Herrington, M., Litovsky, Y., & Gale, H. (2013). *Generation entrepreneur? The state of global youth entrepreneurship* (The Prince's Youth International Business (YBI) and Global Entrepreneurship Monitor (GEM) Report).

Rosa, P. J., Kodithuwakku, S., & Balunywa, W. (2006). Entrepreneurial motivation in developing countries: What does "necessity" and "opportunity" entrepreneurship really mean? *Frontiers of Entrepreneurship Research, 26*, 1–14.

Schoof, U. (2006). *Stimulating youth entrepreneurship: Barriers and incentives to enterprise start-ups by young people* (No. 388157). Geneva: International Labour Organization.

Xavier, S., Kelley, D., Kew, J., Herrington, M., & Vorderwülbecke, A. (2013). *Global Entrepreneurship Monitor (GEM) 2012 Global Report*. GERA/GEM. Retrieved from www.gemconsortium.org/report.

2 Youth entrepreneurship trends and policies in Uganda

Rebecca Namatovu, Thilde Langevang,
Samuel Dawa, and Sarah Kyejjusa

Introduction

Uganda has one of the youngest populations in the world with 78% of the population below the age of 30 (International Youth Foundation [IYF], 2011). While the country has demonstrated high economic growth rates during the last two decades, this growth has not manifested itself in sufficient jobs for the growing youth cohort. Youth unemployment has, therefore, emerged as one of Uganda's major development challenges. In light of the mounting youth unemployment problem and a policy environment centred on the private sector as the primary engine of growth, young Ugandans are increasingly being encouraged by a range of actors to create jobs for themselves and their peers by starting their own enterprises.

Despite this growing interest in youth entrepreneurship, there is only a limited body of knowledge on young Ugandans' entrepreneurial activities and attitudes, and the policy environment surrounding youth entrepreneurship. This chapter aims to help fill this gap by drawing on original data from the nationwide Global Entrepreneurship Monitor (GEM) survey. The chapter also provides an overview of key policies and programmes that focus on promoting youth employment and entrepreneurship in Uganda, and assesses how these correspond to the GEM survey's findings. We begin by presenting the socioeconomic and political environment within which youth entrepreneurship in Uganda is embedded.

Entrepreneurship and youth in Uganda

Entrepreneurship in Uganda is closely interwoven with the country's socioeconomic and political situation (Langevang, Namatovu, & Dawa, 2012). Pre-independence, the Ugandan economy was dominated by Asian business owners. Many of them, however, were expelled from the country in 1972 by the then Ugandan president, Idi Amin, and their businesses allocated to locals, most of whom had few skills or business management experience. This caused many of these businesses – and, consequently, almost the entire economy – to collapse (Ofcansky, 1996). This and other factors triggered civil unrest and related lawlessness, all of which impeded economic activity and entrepreneurship for many years.

By 1987, the economy was plagued by hyperinflation and budgetary deficits amid low levels of productivity in both the agricultural and industrial sectors. Facing a severe economic crisis, the Ugandan government was compelled to undertake the structural adjustment programmes proposed by the World Bank and International Monetary Fund. These involved realigning the exchange rates, liberalising the market, returning expropriated properties, and downsizing public services, among others. The subsequent period witnessed massive layoffs in the public sector, which pushed many Ugandans to business ownership as a source of livelihood (Brett, 1995). However, many of these firms did not survive, and those that did remained very small and informal (Bigsten & Kayizzi-Mugerwa, 1992).

Neoliberal economic policies emphasising free trade and private sector development have continued into the 21st century and Uganda is often highlighted as exemplary when it comes to liberalisation and structural adjustment. Its economic restructuring efforts are widely recognised as having effectively fostered the country's growth (World Bank, 2013). During the last two decades, there has been a marked improvement in key economic indicators with average annual rates of economic growth reaching 7% and poverty levels dropping from 56% in 1992 to 24% in 2010 (World Bank, 2013).

However, recent years have seen increased economic volatility with declining growth rates and high inflation. Moreover, Uganda is still classified as a low-income country and the benefits of growth have not translated directly into measurable improvements in the standard of living for many Ugandans. The growth that has taken place is concentrated in the urban areas of the south-central part of the country. Historically, the Northern Region has been marginalised economically and politically. A two-decade-long civil war fought from 1987 to 2006 between the Ugandan government and the Lord's Resistance Army has only amplified the region's relative underdevelopment (Golooba-Mutebi & Hickey, 2010).

Agriculture remains the cornerstone of the economy, although the sector's contribution to gross domestic product (GDP) has declined over the years. In 2012, it accounted for 24% of GDP compared to 57% in 1990 (World Bank, 2013). About three-quarters of the population engage in agricultural production as their primary livelihood activity. The bulk of this production is subsistence-based rather than commercial, but agricultural products still constitute the majority of Ugandan exports. Coffee remains the main export crop although new crops such as flowers, tobacco, and maize have been introduced in attempts to diversify the economy.

While the agricultural contribution to GDP has decreased, Uganda has a growing services sector (constituting 44% of GDP in 2012) and manufacturing sector (constituting 26% of GDP in 2012) (World Bank, 2013). Although the decreasing role of agriculture is a sign of structural transformation in the economy, this trend has not been followed by adequate job creation: the services and manufacturing sectors record high levels of underemployment (Ahaibwe, Mbowa, & Lwanga, 2013). Enterprises are mostly very small and predominantly found in the informal sector. In 2006, Uganda discovered oil reserves and although these have still not started to generate revenue, oil is expected to contribute significantly to economic growth in the coming years.

Paralleling the country's economic transformations are marked demographic changes. Compared to other African countries, Uganda has relatively low levels of urbanisation, with about 85% of the population living in rural areas. However, the urban population is expected to grow considerably during the next few years. Furthermore, Uganda's population is growing at a rate of 3.2% per annum (Uganda Bureau of Statistics [UBOS], 2010), which means that it is becoming increasingly younger. The growing youth population has, therefore, captured the interest of academics and policy makers alike in recent years. Key to the ongoing discussions is the predicament of young people's employment, given that they comprise the bulk of unemployed persons in Uganda. It is estimated that around 400,000 young people are annually released into the job market to compete for only 9,000 formal jobs available (Bbaale, 2014). This situation has translated into growing unemployment rates: estimates of youth unemployment in Uganda range from 3.8% to as high as 83%, depending on the context and definition of "youth" and "unemployment" (Balunywa et al., 2013; IYF, 2011; Bbaale, 2014). Remaining economically idle in the longer term is not an option for most young people, many of whom find work in the informal sector, in which an estimated 90% of young people are employed (Bbaale, 2014).

Admission rates have increased significantly in primary schools with a net enrolment rate of 90% for both boys and girls. Enrolment rates at secondary schools, however, remain low at 21% (IYF, 2011). Furthermore, the quality and relevance of the education provided is increasingly being questioned, with critics arguing that it is too theoretical and does not equip young people with either employable or entrepreneurial skills. Despite this, as we show in the next section, young people in Uganda are highly entrepreneurial.

Characteristics of youth entrepreneurship in Uganda

This section examines the levels, attitudes, and characteristics of young entrepreneurs. We start by presenting the methodology of the GEM survey in Uganda.

Methodology

The annual household GEM survey for 2012 was administered to 2,343 individuals (aged 18–64) who were randomly selected and thus representative of the Ugandan population. The selection process took place as follows. The survey was conducted in the four geographic regions of Uganda (Western, Eastern, Central, and Northern) and Kampala, which is part of the Central Region but was surveyed separately for the purpose of GEM. Eight districts were randomly selected in each region using the probability proportional to size (pps). A parish was then randomly selected within each district, also based on pps, and within each parish an enumeration area was randomly selected. A designated sample of households was then selected within each enumeration area and one person selected at random per household to be interviewed. The standardised GEM Adult Population Survey questionnaire was used in all of the interviews, which were conducted face-to-face by trained field enumerators who translated the questions into the local languages. Uganda first participated in GEM in 2003 and has since then participated in four other cycles in 2004, 2009, 2010, and 2012. In this chapter, we focus on the 2012 data set and the youth population aged 18–35.

Levels of entrepreneurship

A key measure of entrepreneurial activity used by GEM is total early-stage entrepreneurial activity (TEA), which measures the proportion of the population that owns either a nascent business (less than six months old) or a new business (less than 3.5 years old). Uganda had a TEA of 35.7, which means that, for every 100 adults aged 18–64, 36 were in the process of starting or had recently started a new business the previous year (see Figure 2.1). Youth (aged 18–35) had a significantly higher TEA (39.0) than adults (28.5). With a TEA of 41.1, the age group 25–35 had the highest TEA of all age categories, followed by the 18–24 group, which scored a TEA of 38.4 (see Figure 2.2).

Young women registered slightly lower TEA levels (37.9) than their male counterparts (40.3), but had a higher TEA than older women (31.0). Compared to the general trend for GEM countries (Kelley, Brush, Greene, & Litovsky, 2011), Uganda shows a very high rate of female entrepreneurship for the total population with barely any gender divide between men (36.0) and women (35.5).

Female entrepreneurship is prevalent and increasing in Uganda for a number of reasons. Historically, women were seen as mainly subsistence farmers and homemakers and were not expected to start, own, or manage a business. Nowadays,

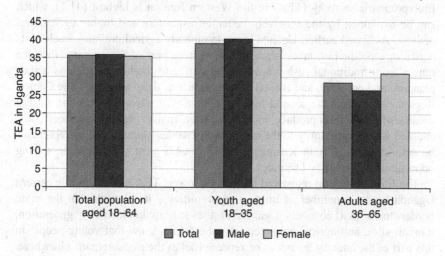

Figure 2.1 TEA in Uganda, by age and gender (source: Uganda GEM survey 2012).

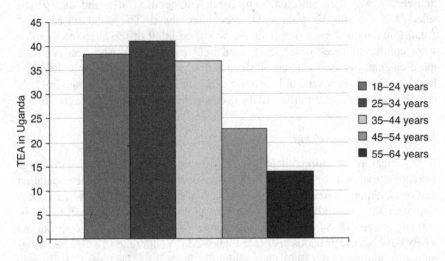

Figure 2.2 TEA in Uganda, by age group (source: Uganda GEM survey 2012).

however, they are increasingly urged to contribute to household welfare (Snyder, 2000) since one income is rarely enough to maintain a family. Women thus engage in a range of income-generating activities, often around the home, such as baking cakes, selling home-grown produce, hairdressing, or offering professional services such as accounting (Bbaale, 2014). Notably, out of every five women who reported "homemaking" as their main occupation, one was engaged in early-stage entrepreneurship. These trends indicate that, increasingly, it is considered perfectly acceptable for women to start up and manage businesses in Uganda.

There are some slight regional differences in the levels of entrepreneurship. Entrepreneurial activity (TEA) in the Western Region is highest (41.1), which can be explained by the relatively better infrastructure and higher agricultural activity. As stated earlier, Uganda is predominantly agricultural and much of its agriculture production takes place in the western part of the country. Produce ranges from traditional cash crops such as coffee to food crops such as maize, plantain, and potatoes, and animal products such as dairy and beef. The Central Region, which had the second highest TEA (39.2), also accounts for a large volume of agricultural produce, including coffee, maize, plantain, and beef. This, coupled with its proximity to the capital city, enables high rates of entrepreneurial activity. The TEA in Kampala was 37.2, with most young people seizing opportunities in trade and services.

The Eastern Region registered the second lowest TEA of 31.2. While eastern Uganda boasts a number of unique opportunities – its proximity to the main border town that connects Uganda to the sea, hydroelectricity generation, tourism sites, and agriculture potential – the results show that young people in this part of the country are not as entrepreneurial as their counterparts elsewhere. Partly, this might be explained by the occurrence in recent years of natural disasters in parts of eastern Uganda, such as landslides in Bududa and flooding in Teso, which have interrupted economic activity. Northern Uganda's TEA of 30.5 was the lowest. Recovering from a two-decade-long insurgency during which young people were some of the main victims of the conflict has severely hindered economic activity among this cohort. Over the past five years, the camps, where the majority of the population was kept, have been closed and substantial government relief efforts and NGO support have been directed at sparking economic activity in the area.

Attitudes towards entrepreneurship

Young Ugandans register very positive attitudes to entrepreneurship, with almost eight out of 10 seeing good opportunities for starting a business within the next six months. About three-quarters of young people who were not already operating a business said that they intended to start one in the near future (see Table 2.1). Furthermore, 88% believed they had the required knowledge to start a business. Notably, 90.3% of all young people said they wanted to be entrepreneurs in the future while only 5.8% expressed a preference for being employed by others. About 18.2%, however, felt that fear of failure might prevent them from starting a business.

Young women score slightly lower on each measure, indicating that they hold somewhat less positive attitudes towards entrepreneurship and have a greater fear of failure. There is no marked difference between youth and adults apart from in their intention to start a business, where young people score higher. This overall optimistic assessment of entrepreneurship by Ugandan youth derives from many sources, including the perceived poor conditions of wage employment, fertile opportunities for entrepreneurship, and a socio-economic and cultural environment that encourages business start-ups (Langevang et al., 2012).

Table 2.1 Attitudes to entrepreneurship in Uganda (%)

Attitudes to entrepreneurship	Youth (18–35)	Male youth (18–35)	Female youth (18–35)	Adults (36–64)
Sees good opportunities for starting a business in the next six months	79.7	81.1	78.5	80.2
Has the required knowledge to start a business	88.0	90.2	86.1	87.5
Fear of failure prevents respondent from starting a business	18.2	15.4	20.5	21.1
Intends to start a business (among non-entrepreneur population)	75.5	79.0	72.4	66.4

Source: Uganda GEM survey (2012).

Profiles of youth entrepreneurs

The data show that most entrepreneurs have less than secondary schooling. A little more than half (52%) the young entrepreneurs surveyed had only completed primary school or less. An additional third (33.5%) had some secondary education, and just 4.3% had completed college or university. These numbers show that the majority of young entrepreneurs in Uganda start their businesses with very limited formal education.

An interesting characteristic of entrepreneurship in Uganda is the prevalence of pluriactivity: the practice of engaging in multiple income-generating activities at once. The self-employed are most likely to set up new enterprises, with 40% of young people already owning a business, intending to start another in six months' time, or already managing another new enterprise. This illustrates a high level of serial entrepreneurship and denotes the practice of continuously engaging in different entrepreneurial activities (Minniti & Naudé, 2010).

Many employed young people also run a business on the side; one in eight engaged in full-time employment are starting or managing a business compared to 7% of those in part-time employment. Earnings from wage employment in Uganda are low, prompting employees to seek alternative means of meeting their needs. It is commonplace for employees to be marketing their wares in the workplace or operating a bar in the evening, for example. Employment also exposes individuals to networks through which they might encounter entrepreneurial ideas, while the wage derived provides capital for start-up. Students also recorded a high TEA of 17%. Those in tertiary institutions in particular tend to engage in business activities in order to pay their tuition fees, sustain themselves, and prepare for life after study, being well aware of the scarcity of jobs (see also Chapter 17 in relation to Ugandan university students). Typical businesses include retail shops, canteens, photocopying services, selling mobile phone airtime, and managing poultry or piggery projects.

With regard to young people's motives for starting a business, slightly more reported being driven by necessity (51.8%) than by opportunity (47.1%), with no significant differences between men and women. These figures indicate that, while many young people perceive good opportunities in a growing economy and on this basis decide to embark on entrepreneurship, somewhat more are compelled to start a business because they perceive no other employment options to be available. It should be highlighted, though, that the motives for starting a business are often complex. Splitting motivation squarely into necessity and opportunity has been increasingly criticised, with studies showing that there may be multiple, interlinked, and shifting motives behind business start-up in Uganda (Langevang et al., 2012; Rosa, Kodithuwakku, & Balunywa, 2006).

Most young entrepreneurs operate in the consumer-oriented sector (80.4%), which is dominated by retail and services (see Table 2.2).[1] Women are pre-eminent here with 87.9% of young female entrepreneurs in this sector compared to 76.9% of young men. Conversely, more men (16.8%) than women (7.6%) operate businesses in the transforming sector. These differences highlight the

Table 2.2 Sector distribution of youth TEA entrepreneurs in Uganda (%)

Sector	Adults (36–64)	Youth (18–35)	Male youth (18–35)	Female youth (18–35)
Extractive Forestry, agriculture, fishing, mining	10.6	5.9	9.1	3.0
Transforming Construction, manufacturing, transportation, communications, utilities, wholesale	11.1	11.9	16.8	7.6
Business services Finance, insurance, real estate, all business services	1.4	1.7	2.0	1.5
Consumer-oriented Retail, motor vehicles, lodging, restaurants, personal services, health, education and social services, recreational services	76.9	80.4	72.1	87.9

Source: Uganda GEM survey (2012).

gendering of businesses: trading is constructed as a key female vocation while construction, manufacturing, and transportation are typically considered male. The data also reveal some generational differences. Notably, more adult entrepreneurs (10.6%) are engaged in the extractive sector (which is dominated by agriculture) compared to youth (5.9%). Conversely, more young entrepreneurs (80.4%) are engaged in the consumer-oriented sector than adults (76.9%).

Although a key way in which entrepreneurship can contribute to economic development is by creating jobs (Minniti & Naudé, 2010), in Uganda the GEM data reveal that entrepreneurship predominantly generates self-employment. Most youth-run businesses have no employees (57.3%) while hardly any (1.5%) have more than six employees and none have more than 20 (see Table 2.3). There are no striking differences between youth and adults, but the data do indicate gender differences. Young women's businesses are smaller, with more young women (61.7%) than men (53.1%) not employing any other people.

Table 2.3 Number of employees of TEA entrepreneurs in Uganda (%)

Number of employees	Youth (18–35)	Male youth (18–35)	Female youth (18–35)	Adults (36–64)
0	57.3	53.1	61.7	54.9
1–5	40.7	44.1	37.9	44.4
6–19	1.5	2.8	0.4	0.7

Source: Uganda GEM survey (2012).

These gender differences are also reflected in the five-year job growth aspirations of young entrepreneurs: 32.8% of female TEA entrepreneurs compared to 24.7% of their male counterparts did not expect to expand their business by employing any additional people (see Table 2.4). Conversely, 13.6% of men expected to build their business by employing 6–19 people, compared to only 6.2% of women, and while 2.1% of men expected to employ more than 20 workers, no women expressed such expectations. These findings correspond to studies showing that women face particular barriers in expanding their businesses (Amine & Staub, 2009; Minniti & Naudé, 2010).

The data also show that, according to their own assessment of the innovativeness of their businesses, relatively few young entrepreneurs (16.1%) consider themselves as offering new products or services. More young men (19.1%) than young women (13.3%) consider their businesses to be innovative. In terms of their assessment of whether there were many other businesses offering the same products or services, two-thirds of the youth surveyed (65.7%) indicated that this was so. These figures imply that competition is high and that replicative entrepreneurship – rather than innovation – characterises young entrepreneurs' undertakings.

Although most young entrepreneurs (78.2%) have no international customers, some young people do trade internationally: 14.3% said that up to one-quarter of their customers were located abroad and 6.7% had more than one-quarter but less than three-quarters located abroad. The prevalence of internationalisation is primarily due to the considerable informal cross-border trade that takes place among the five neighbouring countries (Kenya, Tanzania, Rwanda, the Democratic Republic of the Congo, and South Sudan) (UBOS, 2009).

While young people show high levels of early-stage entrepreneurship, they score remarkably lower on established business ownership, with only 21.5% of the youth surveyed running businesses that are more than 3.5 years old. This reflects the obstacles they face in sustaining their businesses. Such difficulties are more pronounced among women who show lower levels of established business ownership (18.6%) compared to their male peers (24.8%). Furthermore, 20% of the young people had discontinued a business, the most frequently cited reasons for which were that it was not profitable (37.6%) or it was difficult to obtain financing (21.2%).

Table 2.4 Five-year job growth aspirations of TEA entrepreneurs in Uganda (%)

Number of additional employees	Youth (18–35)	Male youth (18–35)	Female youth (18–35)	Adults (36–64)
0	28.9	24.7	32.8	28.6
1–5	60.3	59.6	61.0	59.4
6–19	9.7	13.6	6.2	9.4
20+	1.0	2.1	0	2.6

Source: Uganda GEM survey (2012).

While both youth and adults show similar rates of business discontinuance, a higher percentage of youth (37.6%) compared to adults (25.9%) attributed their business discontinuation to lack of profitability. It should be noted, though, that just because a business fails it does not necessarily imply that the entrepreneur behind it has failed and left his or her entrepreneurial career. As Balunywa et al. (2013) propose, high rates of business discontinuance may be a sign of business dynamism in that some entrepreneurs "churn" businesses by starting new ones and discontinuing those that perform poorly, while trying to start more promising ones.

Policies and programmes focusing on youth entrepreneurship in Uganda

The high rates of youth entrepreneurship and entrepreneurial dynamism in Uganda are paralleled by a growing policy focus on youth entrepreneurship. During the last decade, there has been an upsurge in policies and programmes geared towards young people, and youth employment issues have become an integral part of Uganda's national planning framework.

In the general election held in early 2011, youth employment featured in many presidential candidates' manifestos and was extensively discussed in political debates. The re-elected president of Uganda informed the public of the government's strategies to empower youth and outlined a range of corresponding programmes. Many of the youth initiatives started by the government highlight the importance of having an economically active youth cohort and focus particularly on youth entrepreneurship. Overall, these programmes have taken the shape of capital-based assistance; provision of entrepreneurship training, internships, and apprenticeships; and efforts to recognise and reward successful young entrepreneurs through awards and business plan competitions.

The National Development Plan (2010/2011–2014/2015) underscores the youth unemployment challenge and emphasises the importance of providing young people with quality education and entrepreneurial skills. The interventions outlined focus on fostering youth entrepreneurship and include: establishing start-up business clinics, developing a techno-entrepreneurs' park and small and medium enterprise incubation programme, providing seed capital and motivation to young enterprises, and operationalising regional youth centres and supporting apprenticeship centres (Uganda National Planning Authority, 2010).

The National Youth Policy produced by the government in 2001 is currently being revised, but is expected to emphasise employment creation and to highlight the importance of entrepreneurship (IYF, 2011). The National Employment Policy for Uganda from 2011 also highlights youth employment as a policy priority area. The Skilling Uganda Programme (2011–2020) emanates from an empirical study that exposed how the existing technical and vocational education courses were impractical, did not equip young people with employable skills, and were generally not popular among Ugandans. The programme focuses on business, technical, and vocational education and training and proposes a "paradigm shift" in how education and skills in vocational institutions should be administered and taught. It also seeks to change the general negative social attitudes towards vocational education in Uganda (Uganda Ministry of Education and Sports, 2011).

A key government intervention focusing on youth entrepreneurship is the Youth Venture Capital Fund launched in 2012 to support youth livelihood activity. The programme was designed to give interest-free loans to youth groups of about 10–15 members, but was marred by public criticism from its inception. Specifically, young people felt that the programme was not easily accessible to them because the private banks administering the fund had imposed stringent conditions, such as collateral and formal business registration. In response to this outcry, the government initiated the Youth Livelihood Programme in 2013, which targets out-of-school unemployed youth aged 18–30. Beneficiaries do not need to have collateral to access funds, but they are required to present a three-year work plan and to have a permanent address and group membership of 10–15 (30% of which must be female).

The Peace, Recovery and Development Plan for Northern Uganda (2007–2010) also included a youth entrepreneurship component through the Northern Ugandan Social Action Fund. A key part of this programme was the Youth Opportunities Programme: this provided cash transfers to young people who were encouraged to organise themselves into groups of 15–25 and submit grant proposals for purchasing skills training, tools, and other materials required to start an enterprise. Successful groups received a lump-sum cash transfer and were not subject to supervision. Follow-up surveys of the participants two and four years later found that their incomes had increased significantly. With hundreds of groups applying, however, the demand for the programme far outstripped the supply of funds (Blattman, Fiala, & Martinez, 2014). Furthermore, concerns were raised that the most disadvantaged people – who did not have adequate social and human resources to join a group and/or write a proposal – were excluded from the programme.

Yet another programme focusing on youth entrepreneurship was the Youth Entrepreneurship Facility – an initiative of the Africa Commission, implemented by the Youth Employment Network and the International Labour Organization (ILO) between 2010 and 2014. The facility's mission was to contribute to job creation in East Africa (Uganda, Kenya, and Tanzania) by developing youth entrepreneurship. Its youth-to-youth fund component supported small-scale youth entrepreneurship development projects implemented by youth-led organisations. This included a competitive grant scheme for these organisations to propose innovative project ideas on how to create entrepreneurship and business opportunities for their peers. The organisations with the most innovative project ideas received a grant and basic business training to help them implement their project and test the viability of their ideas. Across the three participating countries, 2,400 applications were submitted by youth groups while only 76 groups received a grant (ILO, 2014).

What emerges from this brief review of policies and programmes is a strong focus on promoting entrepreneurship among youth. Most programmes include elements targeted at helping young people start an enterprise. Financial support is the dominant element, at times combined with skills training. It should be highlighted that the rise of youth entrepreneurship programmes has not led to a concurrent effort to evaluate their impact. There are very few empirically grounded studies that look into the results, strengths, weaknesses, and pitfalls of such interventions. As a result, we know very little about the actual reach or effects of these programmes. However, there is an evident mushrooming of many different programmes with a relatively limited intake and coverage and their collective impact appears minimal. Given the scale of the youth population, there is a real worry that these relatively small, isolated interventions only amount to drops in the ocean.

Conclusion

The GEM data reveal that Ugandan youth are very entrepreneurial and have positive attitudes towards entrepreneurship. Youth-run businesses remain small, however, and are concentrated in sectors of the economy characterised by fierce competition. Consequently, many fail. Young people attribute their business failure to lack of profitability and inadequate access to finance. The data also reflect some gender differences, with young women being slightly less entrepreneurial, operating smaller businesses, and indicating lower growth aspirations than their male peers.

Our review of the main policies geared towards youth employment and entrepreneurship reveals a plethora of programmes focused on promoting start-up activity among youth – predominantly through financial assistance and, to a lesser extent, skills training. The absence of thorough impact studies makes it impossible to conclude on the consequences of such programmes, but their intake is fairly limited and their impact, therefore, appears minimal. Furthermore, the focus on financial support to business start-up seems too narrow when juxtaposed with the GEM results, which reveal that a considerable number of young people are already engaged in early-stage business activity.

The data show that the major challenge facing youth is not so much starting a business as sustaining and expanding it. Considering that the high discontinuance rate was attributed to lack of access to funds and non-profitability of ventures, it is imperative that interventions also cater to the short-term financial challenges of existing entrepreneurs. This requires finding innovative ways of including youth in the provision of financial services. Given the strong concentration of young people in heavily competitive sectors, it may be wise to develop sector-specific programmes that help them sustain their businesses in these sectors, become more innovative, and enter other sectors of the economy. While some programmes include education and training, there also seems to be scope for strengthening this element, taking into account the limited educational background of most Ugandan entrepreneurs. Furthermore, although the gender differences in entrepreneurship in Uganda are less marked than in many other countries, there is still a need to devise interventions that address gender-related barriers to business start-up and growth.

Note

1 The classification into extractive, transforming, business services, and consumer-oriented sectors is devised by GEM, adopted from the World Economic Forum.

References

Ahaibwe, G., Mbowa, S., & Lwanga, M. M. (2013). *Youth engagement in agriculture in Uganda: Challenges and prospects* (Research Series No. 106). Kampala: Economic Policy Research Centre.

Amine, L. S., & Staub, K. M. (2009). Women entrepreneurs in sub-Saharan Africa: An institutional theory analysis from a social marketing point of view. *Entrepreneurship & Regional Development: An International Journal, 21*(2), 183–211.

Balunywa, W., Rosa, P., Dawa, S., Namatovu, R., Kyejjusa, S., & Ntamu, D. (2013). *Global entrepreneurship monitor: GEM Uganda 2012 executive report*. Kampala: Makerere University Business School and International Development Research Centre.

Bbaale, E. (2014). Where are the Ugandan youth? Socioeconomic characteristics and implications for youth employment in Uganda. *Journal of Politics and Law, 7*(1), 37–63.

Bigsten, A., & Kayizzi-Mugerwa, S. (1992). Adaptation and distress in the urban economy: A study of Kampala households. *World Development, 20*(10), 1423–1441.

Blattman, C., Fiala, N., & Martinez, S. (2014). Generating skilled self-employment in developing countries: Experimental evidence from Uganda. *Quarterly Journal of Economics, 129*(2), 697–752.

Brett, E. A. (1995). *Structural adjustment in Uganda: 1987–1994*. Copenhagen: Centre for Development Research.

Golooba-Mutebi, F., & Hickey, S. (2010). Governing chronic poverty under inclusive liberalism: The case of the Northern Uganda Social Action Fund. *Journal of Development Studies, 46*(7), 1216–1239.

International Labour Organization. (2014). *The Youth Entrepreneurship Facility's Youth-to-Youth Fund in East Africa: Public–private partnership* [Fact sheet]. Retrieved from www.ilo.org/pardev/public-private-partnerships/WCMS_193814/lang-en/index.htm.

International Youth Foundation (IYF). (2011). *Navigating challenges. Charting hope: A cross-sector situational analysis on youth in Uganda* (Vol. 1). Kampala: International Youth Foundation.

Kelley, D. J., Brush, C. G., Greene, P. G., & Litovsky, Y. (2011). *Global entrepreneurship monitor: 2010 women's report*. Babson Park, MA: Babson College, Centre for Women's Entrepreneurial Leadership. Retrieved from www.gemconsortium.org/docs/download/768.

Langevang, T., Namatovu, R., & Dawa, S. (2012). Beyond necessity and opportunity entrepreneurship: Motivations and aspirations of young entrepreneurs in Uganda. *International Development Planning Review, 34*(4), 439–460.

Minniti, M., & Naudé, W. A. (2010). What do we know about the patterns and determinants of female entrepreneurship across countries? *European Journal of Development Research, 22*(3), 277–293.

Ofcansky, T. P. (1996). *Uganda: Tarnished pearl of Africa*. Boulder, CO: Westview Press.

Rosa, P. J., Kodithuwakku, S., & Balunywa, W. (2006). Entrepreneurial motivation in developing countries: What does "necessity" and "opportunity" entrepreneurship really mean? *Frontiers of Entrepreneurship Research, 26*(20), 1–13.

Snyder, M. (2000). *Women in African economies: From burning sun to boardroom.* Kampala: Fountain Publishers.

Uganda Bureau of Statistics (UBOS). (2006). *The 2002 Uganda population and housing census: Gender and special interest groups.* Kampala: Uganda Bureau of Statistics.

Uganda Bureau of Statistics (UBOS). (2009). *The informal cross-border trade qualitative baseline study 2008.* Kampala: Uganda Bureau of Statistics.

Uganda Bureau of Statistics (UBOS). (2010). *Statistical abstract.* Kampala: Uganda Bureau of Statistics.

Uganda Ministry of Education and Sports. (2011). *Skilling Uganda: BTVET Strategic Plan (2011–2020).* Kampala: Uganda Ministry of Education and Sports.

Uganda National Planning Authority. (2010). *National Development Plan (2010/11–2014/15).* Kampala: Uganda National Planning Authority.

World Bank. (2013). *Uganda economic update: Bridges across borders – Unleashing Uganda's regional trade potential.* Washington, DC: World Bank.

3 Youth entrepreneurship in Ghana

Current trends and policies

*George Owusu, Paul W. K. Yankson, and
Robert Darko Osei*

Introduction

Economic and political reforms in Ghana over the last three decades have resulted in a stable socioeconomic environment and generated significant economic growth rates with a strong focus on private sector development. These growth rates, however, have not yielded the desired level of employment in the private sector, which is dominated by the informal sector. The country's relatively high economic growth rates have been driven largely by the substantial expansion in nonmanufacturing sectors, such as mining and other extractive industries, which are capital-intensive and generate relatively few job opportunities (Cornelius, Landström & Persson, 2006). Again, the liberalisation and privatisation of the economy have been accompanied by declining employment levels in the public sector. The overall result is decreasing employment opportunities in the formal sector and a swelling number of workers in the informal sector (Langevang, 2008).

This chapter presents the overall socioeconomic environment within which youth entrepreneurship in Ghana is embedded, examines the levels and characteristics of youth entrepreneurship across the country, and provides an overview of key policies and interventions promoting youth employment and entrepreneurship. To overcome the challenges of the paucity of data on youth entrepreneurship in Ghana, we draw heavily on the 2012 Global Entrepreneurship Monitor (GEM) survey conducted across all regions of Ghana by the Institute of Statistical, Social and Economic Research (ISSER) at the University of Ghana.

Brief overview of Ghana's economy and society

Ghana was the first sub-Saharan African country to gain independence from colonial rule in 1957 with the hope of promoting rapid economic growth and development on the African continent. While Ghana experienced reasonably high growth soon after independence, by 1965 the country's per capita growth rate was already negative. Economic conditions appeared to improve significantly during the late 1960s and the early 1970s but the mid-1970s saw the beginning of significant deterioration due to mismanagement and political instability (Fosu, Aryeetey, & Quartey, in press). This drastic economic decline persisted and, by the early 1980s, Ghana like many sub-Saharan African countries was forced to seek the support of the World Bank and International Monetary Fund's Structural Adjustment Programme and Economic Recovery Programme interventions.

Although economic conditions in Ghana have improved significantly since the mid-1980s, it was not until 2003 that per capita income surpassed its 1960 level (Fosu et al., in press). In addition, the level of poverty fell from 51.7% in 1991 to 39.5% in 1999 to 28.5% in 2006 and further to 24.2% in 2013 (Ghana Statistical Service, 2007, 2014). Much of Ghana's economic growth and poverty reduction have been driven by foreign direct investment in its natural resources, particularly gold and, in more recent years, oil production (Owusu & Afutu-Kotey, 2014). These investments appear to have boosted the services sector and with the continuous contribution of the agricultural sector, particularly the export of cocoa and other cash crops, Ghana was declared a lower middle-income country in 2010.

Despite improvements in the performance of the Ghanaian economy over a relatively long period, its structure remains largely unchanged. In addition, the benefits of economic growth have not been uniform across regions: poverty is still high in rural Ghana, especially in the northern parts of the country (the Upper East, Upper West, and Northern Regions) which, according to the Ghana Statistical Service (2014), account for 40% of overall poverty in the country. This situation has limited the economy's capacity to achieve sustainable improvements in livelihoods and a broad-based development that benefits the bulk of the population. Indeed, it is clear that, despite government efforts since independence, the structure of the Ghanaian economy has not been significantly transformed to allow the population to earn higher incomes from high-value activities, thus limiting considerably the capacity of individuals and households to participate effectively in the economy. In turn, this has led to unemployment, underemployment, and disguised unemployment.

Although the contribution of agriculture to the national GDP has declined over the years, the sector remains the biggest employer of the economically active population aged 15 years and above. The sector's contribution to GDP has declined from about 50% in 1965 to 28% in 2007 and then to 22% in 2013 (Dzanku & Aidam, 2013; ISSER, 2014). As of 2010, about 42% of Ghana's economically active population was employed in agriculture; in rural areas, this proportion is as high as 56.8% and 73.1% among female- and male-headed households, respectively. Despite the declining agricultural contribution to GDP relative to the services and industry sectors, it is still difficult to talk about the structural change of the Ghanaian economy, given the large presence of labour and continuous use of rudimentary production methods in the sector (Dzanku & Aidam, 2013). Again, the growth recorded in the services and industry sectors has not been backed by adequate job creation. Their contribution is expected to continue to grow in coming years and decades, especially following the discovery of oil in 2007 and the start of production in 2010.

Accompanying the economic growth of recent decades are marked demographic changes. Compared to other sub-Saharan African countries, Ghana can be described as relatively well urbanised with about 51% of its total population living in towns and cities in 2010; the continuous shift of the population from rural to urban centres is unlikely to slow down until 2030 (Owusu & Oteng-Ababio, 2015). The distribution of the urban population, however, is skewed, with a greater concentration in centres such as Accra and Kumasi and to some extent in other secondary centres. This is largely a result of migrants' preference for large urban centres, which offer better employment and other socioeconomic prospects. The movement to cities is dominated by youth, mainly basic and secondary school leavers as well as others who have lost interest in agriculture and rural life in general (Awumbila, Owusu, & Teye, 2014). The large number of unemployed and underemployed youth in urban centres in recent times has captured the interest of academics and policy makers alike. Even though employment data on Ghana is difficult to come by (see Baah-Boateng & Turkson, 2005; Ghana Trades Union Congress, 2005; ISSER, 2010, 2012), existing studies indicate that the population cohort most seriously affected by unemployment and underemployment is youth, with the 20–25-year-old age group being worst hit (Baah & Otoo, 2006).

A combination of factors accounts for the high incidence of unemployment among youth in Ghana (ISSER, 2010). These include: school-to-work transition failures (especially among junior and senior high school graduates who are unable to proceed to the tertiary level but are not adequately prepared with the basic skills needed to succeed in the workplace); the mismatch between labour demand and supply even within the technical and vocational education and training (TVET) system; and inadequate enabling macro-policies that promote youth entrepreneurship and employment. Even though school enrolment across all education levels has improved significantly for both males and females, questions have been raised regarding the quality and relevance of the education and employability of school leavers.

Characteristics of youth entrepreneurship in Ghana

While the literature on entrepreneurship is now vast, until quite recently little was known about the forms and characteristics of entrepreneurship in sub-Saharan Africa in general and youth entrepreneurship in particular. This has improved in recent years with the increasing number of African countries participating in the GEM survey. In this section, we draw on the 2012 GEM survey to examine the characteristics of youth entrepreneurship in Ghana by first providing the methodology for the data collected.

Methodology

The 2012 GEM Ghana survey was an individual-level study with the universe comprising all individuals aged 18–64 living in identifiable residential premises as defined by the Ghana Statistical Service. This excludes institutional populations (people in hospitals, prisons, hotels, boarding schools, temporary camps, and other similar establishments) as well as areas determined to be inaccessible, such as game and forest reserves. Thus, except for age, no other demographic variable was used as a basis for selection, and variables such as ethnicity, religion, and sex were randomly determined at the point of interview.

In accordance with GEM sample size standards, a minimum sample of 2,000 was considered adequate to capture the level of entrepreneurial activity at the national level. A sample size of 2,222 was used, making room for a possible non-response rate of 10% and to improve on the precision of estimates if the non-response rate was less than 10%. To achieve a nationally representative sample in terms of geographical spread as defined by the 10 administrative regions, a multi-stage stratified sampling design was employed. The first stage of stratification involved splitting the population by region of residence. The desired sample size of individuals was allocated to each region in proportion to its population as indicated by the regional distribution of the population from the results of the 2010 Ghana Population and Housing Census. The appropriate number of districts and enumeration areas for each region and district, respectively, was then randomly selected. In all, the survey covered 147 enumeration areas in 44 districts across the regions constituting Ghana.

Levels of entrepreneurship

A key measure of the GEM survey is total entrepreneurial activity (TEA), which measures the participation of individuals in early-stage entrepreneurial activity expressed as a percentage of the adult population (18–64 years) that is in the process of starting or has recently started a business. Figure 3.1 compares TEA rates for males and females for the total population aged 18–64 and their distribution

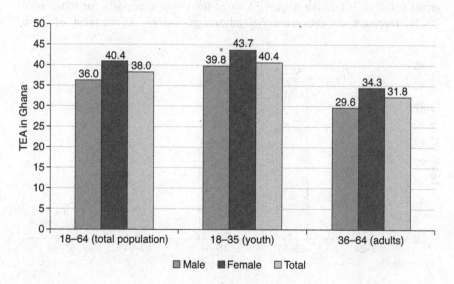

Figure 3.1 TEA in Ghana, by gender (source: Ghana GEM survey 2012).

between adults (36–64 years) and youth (18–35 years) for Ghana in 2012. Across all categories, females tend to have the highest TEA rates relative to their male counterparts. Indeed, the female TEA rate is highest among the youth cohort – almost 44% compared to about 40% for males.

An analysis of past GEM survey results indicates that Ghana is among the very few countries in the world where women appear to be more entrepreneurial than men. This is partly attributable to factors such as the cultural acceptability of women running businesses, growing unemployment and underemployment among their male partners, and the inability of male partners to meet the increasing cost of living (Langevang, Gough, Yankson, Owusu, & Osei, 2015). Consequently, the income women earn through their businesses and income-generating activities becomes indispensable to household survival (Dovi, 2006; Dzisi, 2008; Owusu & Lund, 2004; Robertson, 1995; Yankson, Owusu, Gough, Osei, & Langevang, 2011). Dzisi (2008) argues that the growth of women entrepreneurs needs to be regarded as part of a broader process of social change that is marked by an increasing number of women in the workforce – a process associated with more education for women, postponement of early marriage, smaller families, and an increased desire for financial independence. She adds that, in some contexts, entrepreneurship is an accepted career path for women besides their reproductive role because it is seen as offering employment flexibility and independence, which typical formal employment does not.

Further disaggregating the TEA for Ghana for 2012 by age group shows that youth aged 25–34 represent the most entrepreneurial cohort. As Figure 3.2 shows, for 100 Ghanaians surveyed who were in the process of starting or had recently started a new business in the last year, almost 42% fell within the 25–34-year-old demographic. This is followed by the 35–44 age group (22.5%) and 18–24 age group (21.4%). While the high TEA rates for youth, especially for those aged 25–34, represent an entrepreneurial advantage, further development of these

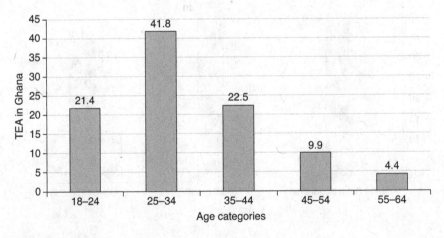

Figure 3.2 TEA in Ghana, by age group (source: Ghana GEM survey 2012).

individuals depends largely on education and the extent to which both public and private initiatives can be fashioned to support them. This point is critical because disaggregating the GEM data on Ghana reveals a high incidence of business closure or discontinuation.

Indeed in 2010, Ghana's rate of business discontinuation – 26% – was the second highest after Uganda among the GEM countries. This declined to 16% in 2012 – the fifth highest among the GEM countries. In other words, despite the decline, the rate of business discontinuation remained high. Yankson et al. (2011) argue that various reasons explain the high rate of business discontinuation in Ghana, including limited profitability, finance, and business opportunities as well as the near absence of innovative training and a regulatory environment that is not conducive to entrepreneurial activity. These factors tend to weigh heavily against young people venturing into the world of business.

Attitudes towards entrepreneurship

It should be stressed that attitudes towards youth entrepreneurship are embedded in several factors, including the sociocultural legitimacy and acceptance of young entrepreneurs; entrepreneurship education and training; access to finance and start-up capital; risk and self-confidence; rigorous administrative and regulatory frameworks; business development and support services; and the overall performance of the economy (Krumina & Paalzow, 2012). In a nutshell, while these factors form the key deficiencies, constraints, and impediments that young people face, they also constitute the basis for developing incentives, strategies, and tools that make – or could make – starting a business a more viable alternative for youth (Schoof, 2006).

Despite the odds, many young Ghanaians continue to set up their own enterprises, particularly in the informal sector (Langevang, 2008). Table 3.1 reveals that Ghana's population has a strong, positive attitude to a number of indicators of entrepreneurship. However, these ratings tend to be relatively high for youth compared to adults, with the exception of "sees good opportunities for starting a business in the next 6 months" and "has the required knowledge to start a business". Again, a closer look at Table 3.1 shows that male youth rate higher on all the indicators than their female counterparts.

The perception that people have the required knowledge, skills, and experience to set up a business may be influenced in part by factors such as unemployment, which can force someone into engaging in a business when the circumstances so demand. Nevertheless, the relatively high proportion of Ghanaian youth that intend to start a business as well as their overall positive attitude to entrepreneurship illustrate young people's drive to engage in entrepreneurial activities – energies that, if supported, could lead to job creation, economic growth, and improved well-being both for the youth and general population.

Table 3.1 Attitudes to entrepreneurship in Ghana (%)

Attitudes to entrepreneurship	Youth (18–35)			Adults (36–64)
	All	Male	Female	
Sees good opportunities for starting a business in the next six months	72.8	73.9	71.7	73.6
Has the required knowledge to start a business	86.3	89.1	83.2	87.5
Fear of failure would prevent respondent from starting the business	20.0	20.4	19.6	14.9
Intends to start a business (among non-entrepreneur population)	61.1	62.1	60.0	53.1
Most people consider starting a business a desirable career option	80.3	82.0	78.5	79.6
Successful entrepreneurs enjoy a high level of status and respect	89.0	89.8	88.1	88.4
There are stories in the public media about successful new businesses	81.9	84.8	78.7	78.4

Source: Ghana GEM survey (2012).

Profiles of youth entrepreneurs

In terms of education, about 41.4% of youth had completed junior high school or less in 2012, while almost 9.6% had no formal education. Almost 24% had secondary education, with only 4.3% having completed university. The overall picture indicates a limited level of educational attainment among youth: about one in 10 have no formal education and even school graduates' competencies in literacy and numeracy – and their employability – is questionable. This limited education among a large section of youth is a major challenge to youth entrepreneurship development in Ghana (ISSER, 2012). While, arguably, many factors contribute to entrepreneurial success, one easily influenced determinant of entrepreneur outcomes is education. Again, it has been proven that education leads to a higher quality of entrepreneurial performance, which should justify investment in the education of (prospective) entrepreneurs (ISSER, 2012).

Entrepreneurship, according to GEM, can be classified by sector: the extractive sector (farming, forestry, fishing, and mining); the transforming sector (manufacturing, construction, transportation, communication, utilities, and wholesale); business services (finance, insurance, and real estate); and consumer services (retailing, lodging, restaurants, personal services, health, education and social services, and recreational services). Similar to the general GEM adult population survey in Ghana, a large proportion of youth operate in the consumer services sector, which involves mainly petty trading and retail in the informal sector (Table 3.2). This is followed by the extractive and transforming sectors.

Table 3.2 Sector distribution of TEA entrepreneurs in Ghana (%)

Sectors of business activity	Youth (18–35)			Adults (36–64)
	All	*Male*	*Female*	
Extractive (forestry, agriculture, fishing, mining)	19.1	25.7	12.5	32.6
Transforming (construction, manufacturing, transportation, communication, utilities, wholesale)	14.1	18.2	10.0	9.2
Business services (finance, insurance, real estate, all business services)	0.9	1.5	0.4	0.4
Consumer-oriented (retail, motor vehicles, lodging, restaurants, personal services, health, education and social services, recreational services)	65.9	54.6	77.1	57.7

Source: Ghana GEM survey (2012).

Disaggregating the data on the sectors in which young people operate reveals that, although a large proportion of youth work in consumer services, more young women than men operate in this sector (Table 3.2). Table 3.2 indicates that almost eight in 10 female youth entrepreneurs can be found in the consumer services sector compared to about five in 10 for their male counterparts. On the other hand, about twice as many male youth (25.7%) operate in the extractive sector compared to 12.5% of female youth. The strong presence of male and female youth in different sectors is partly an outcome of interrelated factors such as education, differential access and rights to resources, and socially gendered roles. For instance, petty trading, the largest economic activity in consumer services, is largely regarded in Ghanaian society as a female activity. On the other hand, women's relatively limited access to land and other land-based resources restricts their presence in the extractive sector, explaining why young men dominate this sector.

A key argument for promoting entrepreneurship is the view that it helps create much-needed jobs for youth and other jobseekers. The Ghana GEM data reveal that entrepreneurship predominantly generates self-employment; Table 3.3 shows that most TEA entrepreneurs have no employees. This is particularly so for youth and, to a greater degree, for young women where as many as almost 74% have no employees. The lack of employees is due to a number of factors, primarily the small size of the business or scale of operation and the weak business environment (e.g. limited credit and high interest rates); these limit investment and business expansion (Langevang & Gough, 2012). Despite these constraints, Table 3.3 shows that a substantial proportion of youth entrepreneurs employ between one and five employees, and 9.3% of male youth have six to 20 employees. The table reveals that twice as many male youth as adults (36–64 years) tend to have six to 20 employees.

Table 3.3 Number of employees of TEA entrepreneurs in Ghana (%)

Number of employees	Youth (18–35)	Male youth (18–35)	Female youth (18–35)	Adults (36–64)
0	52.2	37.0	73.7	49.3
1–5	42.4	53.7	26.3	46.4
6–20	5.4	9.3	–	4.3

Source: Ghana GEM survey (2012).

Gender differences are also reflected in the five-year job growth aspirations of young entrepreneurs: 53.7% of female TEA entrepreneurs compared to 35.2% of their male counterparts do not expect to expand their business by employing any additional people (Table 3.4). The table shows that a relatively high proportion of young male TEA entrepreneurs aspire to build up their businesses and take on more employees than their female counterparts.

While the youth TEA rate tends to be very high, fear of business failure is relatively high among youth, especially the 25–34 entrepreneurial age group. This declines sharply with age (after the mid-40s), however, possibly reflecting stronger social networks, experience and maturity, and a better financial position. Some analysts argue, however, that business discontinuation need not be seen in negative terms at all times; high rates of business discontinuance may reflect conditions of business dynamism in which businesses are created, non-performing ones are discontinued, and new vibrant ones are set up by the same entrepreneurs and/or new entrepreneurs (Rosa et al., 2014). Xavier, Kelley, Herrington, and Vorderwülbecke (2013) argue that even where entrepreneurs (including youth) move on to become employees of other entrepreneurs after abandoning their initial venture, this does not necessarily represent failure because they are likely to avail their experience at their new place of work, with a positive impact on their long-term career path.

Youth entrepreneurship policies and programmes in Ghana

Even though Ghana lacks an explicitly defined youth entrepreneurship policy, two factors have forced policy makers to pay attention to youth entrepreneurship, albeit

Table 3.4 Five-year job growth aspirations of TEA entrepreneurs in Ghana (%)

Number of additional employees	Youth (18–35)	Male youth (18–35)	Female youth (18–35)	Adults (36–64)
0	44.5	35.2	53.7	45.0
1–5	41.1	44.6	37.7	38.7
6–19	11.8	15.7	7.8	11.8
20+	2.6	4.5	0.8	4.6

Source: Ghana GEM survey (2012).

in an uncoordinated manner. These include: persistently high levels of unemployment and underemployment among school leavers, dropouts, and those who have never attended school; and increasing streetism, youth criminal activity, and young people's tendency to channel their energies into anti-social activities (Ghana National Development Planning Commission, 2010). As such, past national medium-term development policy frameworks – the Ghana Poverty Reduction Strategy I (2002–2004), the Growth and Poverty Reduction Strategy II (2005–2009), and the Ghana Shared Growth Development Agenda (GSGDA) (2010–2013) – have all devoted some level of attention to youth employment and entrepreneurship.

The state's increasing attention to youth development culminated in the formulation and promulgation of the first ever National Youth Policy (NYP). Passed and launched in August 2010, the NYP has a long and contentious history. It took over 10 years for the policy to finally see the light of day. Themed "Towards an empowered youth impacting positively on national development", it is intended to provide guidelines for all stakeholders involved in implementing policies, programmes, and projects for the development of youth.

The NYP provides a detailed list of the challenges confronting youth in Ghana: inadequate access to quality education, unemployment and underemployment resulting from inadequate and inappropriate training, the erosion of traditional social support systems for young persons and the weakened role of the family (leading to deviance), and inadequate mentoring opportunities. Against this backdrop, the policy pinpoints 19 priority areas for action. In the area of youth employment and entrepreneurship, the following priorities are most relevant:

1 *Education and skills training*: The policy notes that education and skills training are critical to the development of a young person's productive and responsible life. The main goal of the policy is to "ensure the development of a knowledgeable, self-reliant, skilled, disciplined, and a healthy population with the capacity to drive and sustain the socioeconomic development of the nation" (Ministry of Youth and Sports, 2010: 10). The key consideration here is to increase education and skills training.
2 *Youth and employment*: The policy seeks to build the capacity of youth to discover wealth-creating opportunities in their local environments, enhance access to reliable and adequate labour market information, create opportunities for young people to take advantage of available jobs, and train and prepare youth for the global market.
3 *Entrepreneurial development*: The NYP notes that entrepreneurial development, among other factors, propels and accelerates socioeconomic development. This is limited, however, to a small section of youth. The policy seeks to promote entrepreneurial development by: mainstreaming it into school curricula and integrating entrepreneurial skills into youth development activities; facilitating access to credit for youth; creating corps of young entrepreneurs to serve as role models; and celebrating successful young entrepreneurs.

4   *Youth in modern agriculture*: The goal of the policy here is to promote
    youth participation in agriculture through the following policy objectives:
    promoting their participation in modern agriculture as a viable career oppor-
    tunity and as an economic and business option, and providing resources for
    their participation in modern agriculture.

In terms of implementation mechanisms, the policy states that the National
Youth Council will facilitate and institute a youth stakeholders' forum. This
forum, together with the council and Ministry of Youth and Sports, will oversee
the implementation of the policy. The NYP also calls for an action plan outlining
comprehensive strategies, projects and activities, and the time frames and budget
lines necessary to achieve the desired objectives.

Box 3.1 illustrates that, beyond broad policy measures, there have been
attempts by the government, private sector, and civil society organisations at
both national and local levels to initiate specific programmes and projects promoting youth entrepreneurship. However, many analysts argue that these programmes are not properly evaluated and, in many cases, have had limited
success due to a number of challenges, including lack of demand for the particular types of skills being taught and the poor quality of training provided
(ISSER, 2010). Moreover, the employment outcomes remain unknown; in many
cases, programme graduates are not faring well and/or have remained unemployed since leaving the programme (Palmer, 2006). More importantly, recent
investigations by the media and state have exposed some youth-specific interventions as being plagued with corruption and a means of siphoning scarce state
resources. Of particular interest here is the Ghana Youth Employment and Entrepreneurial Development Agency (GYEEDA), formerly the National Youth
Employment Programme – one of the biggest interventions by the state to deal
with youth unemployment and promote youth entrepreneurship. A recent report
by the Ministry of Youth and Sports (2013) uncovered corruption and irregulari-
ties, including expenditure of several million Ghana cedis on dubious projects,
violations of the state's procurement laws, and suspicious contracts with private
companies, etc.

Box 3.1 Selected youth entrepreneurship programmes and projects in Ghana

Government/public initiatives

- GYEEDA has many modules promoting youth entrepreneurship, for instance, a youth agri-business module. While it focused initially on employing youth directly in various public sector agencies, recent efforts have been directed at promoting entrepreneurship and producing self-employment.
- The Local Enterprise and Skills Development Programme was established in 2011 to provide unemployed youth with basic training in business and entre-preneurial management skills. It targeted 20,000 participants drawn from all

districts across Ghana by 2012. The programme's long-term goals include: equipping beneficiaries with essential capabilities with which to effectively manage their enterprises, enhancing the socioeconomic conditions of youth, and developing the informal sector to contribute optimally to national development.

- The Council for Technical and Vocational Educational and Training has the key goal of ensuring that the unemployed, particularly among youth, are given competitive, employable, and entrepreneurial skills within the formal and informal sectors. It also ensures that graduates of formal, informal, and non-formal TVET institutions are endowed with employable and entrepreneurial skills.
- Business advisory centres and entrepreneurship and enterprise development programmes provide tailor-made training for existing and start-up entrepreneurs (including youth) in three main areas: small business management training, technical training, and entrepreneurship (entrepreneurship awareness seminars and development programmes, and start-your-own-business programmes).
- The National Board for Small Scale Industries provides support to micro and small enterprises (including those run by youth) as well as entrepreneurship development.
- Integrated community centres for employable skills (currently 91 across the country) provide training in various trades to youth over a two- or three-year span. The aim of this training is to produce self-employment.
- Opportunities industrialisation centres provide training to disadvantaged youth, mainly junior and senior high school dropouts and graduates. Graduates include both non-literate and semi-literate youth. These centres can be found in Accra, Kumasi, and Sekondi-Takoradi.

Private sector and NGO initiatives

- TechnoServe's "Believe Begin Become" business plan competition was developed to support and build new businesses by providing training, networking opportunities, and access to capital for business expansion.
- The African Young Entrepreneurs business plan competition – an initiative of the international students' organisation, the Association Internationale des Etudiants en Sciences Economiques et Commerciales – promotes youth entrepreneurship as an important source of innovation and creativity, a solution to unemployment, and a catalyst for socioeconomic development in Africa.
- Under the Junior Achievers Trust International, students starting businesses in school programmes work with senior high school youth. This includes entrepreneurial training and insurance to cushion students against the possible death of their parents.
- EMPRETEC's Personal Entrepreneurial Competencies programme deals with "risk taking and opportunity seeking" and involves start-up modules in human resource management, effective sales strategy, financial management, productivity management, business survival, and growth.

Multilateral and bilateral development organisations' initiatives

- The Youth Employment Network was established in 2001 as an alliance between the United Nations, the International Labour Organization, and the

World Bank to develop and implement strategies that would give young people everywhere a chance to find decent and productive work. The focus of the initiative includes employment creation, employability, entrepreneurship, and equal opportunities for youth.

• The World Bank's support to Ghana is guided by its Country Assistance Strategy, which is aligned with Ghana's medium-term development framework, the GSGDA (2010–2013). This operates through three windows that directly and indirectly support youth entrepreneurship and employment issues: education; medium, small, and micro-enterprise development; and budget support.

Despite criticisms of past and ongoing youth entrepreneurship programmes and projects, in August 2014 the government established yet another state-run programme, the Youth Enterprise Support (YES) initiative. YES has a seed fund of GH₵10 million (about US$3.3 million) to assist young Ghanaians with creative and innovative business ideas and plans to achieve their full potential. The intervention is expected to operate as a multi-sector initiative under the Office of the President, drawing on collaborative support from key agencies such as the National Board for Small Scale Industries, the National Youth Authority, the Ministry of Youth and Sports, the Ministry of Trade and Industry, and the Ministry of Finance.

To apply for YES support, prospective young entrepreneurs must go through the following process (YES, 2015):

• Dates for the submission of business proposals/plans are set and announced by the YES secretariat. Proposals can be submitted via email, post, or an online form or submitted directly to the secretariat.
• A team of YES business consultants shortlists the proposals/plans received. The owners of the selected business proposals/plans for the next stage are then announced.
• An entrepreneurship business clinic is held for all selected business proposal/plan owners. The clinic focuses on helping entrants to streamline their business proposals/plans.
• After the clinic, the selected business owners are asked to revise and resubmit their business proposals/plans.
• The selected business owners are then invited to a personal presentation session.
• A final shortlist of YES beneficiaries or projects to be supported is announced after the session.

The call for YES business proposals was announced in September 2014 and the process of application, shortlisting, training, presentation of revised proposals, and awarding support was expected to finish in December 2014. However, YES has already attracted similar criticism to earlier government initiatives, a key critique being that it duplicates efforts and merely establishes another state bur-eaucracy. More importantly, the sustainability of the programme has been questioned: much of the criticism comes from the political opposition, which sees the programme largely as a creation of the ruling government and a waste of scarce state funds, especially given the number of existing state agencies with the same mandate as YES. In other words, YES is seen in political party colours; should there be a change of government in the future, the continued existence of the programme would be in doubt.

Conclusion

The long period of stable economic growth experienced by Ghana in recent decades has failed to translate into the creation of sufficient employment, especially for young people entering the labour market or the world of work. Nevertheless, the 2012 Ghana GEM survey reveals that Ghanaian youth are highly entrepreneurial and have positive attitudes towards entrepreneurship. Through entrepreneurship they are able to create jobs for themselves as well as employ others. They operate at a small scale, however, due to a range of factors, including limited and high credit costs, limited education and experience, and limited state support.

Given the great potential of youth entrepreneurship, the past few years have seen the government, private sector, NGOs, and international donor agencies introduce several youth entrepreneurship programmes. In particular, the NYP has passed into law and its medium-term development policy framework emphasises youth employment and entrepreneurship. The challenges of youth entrepreneurship, however, cannot be abstracted from the situation of the informal sector in particular and private sector development in general. The view here is that any policy approach that focuses narrowly on promoting "youth entrepreneurship" without attempting to address the broader constraints affecting the informal sector or entrepreneurship in general is not likely to succeed. Again, supporting entrepreneurship should not be celebrated as the only way of generating youth employment, but rather be viewed as one element of a comprehensive employment and development policy that addresses the complex factors and relationships affecting young people's access to meaningful employment.

In conclusion, youth entrepreneurship offers Ghana an immense opportunity to accelerate youth development in terms of capital, employment training, and livelihood opportunities. Policies designed to promote youth entrepreneurship should ensure that these young entrepreneurs have access to cheaper sources of capital, training, land, and other enabling factors. Youth entrepreneurship can only serve as a catalyst to development if these enabling factors are made available.

References

Awumbila, M., Owusu, G., & Teye, J. K. (2014). *Can rural–urban migration into slums reduce poverty? Evidence from Ghana* (Migrating out of Poverty Working Paper No. 13). London: DfID.

Baah, Y. A., & Otoo, I. (2006). *Earnings in the private formal and informal economies in Ghana*. Accra: Ghana Trades Union Congress.

Baah-Boateng, W., & Turkson, F. B. (2005). Employment. In E. Aryeetey (Ed.), *Globalization, employment and poverty reduction: A case study of Ghana* (pp. 104–139). Accra: Institute of Statistical, Social and Economic Research.

Cornelius, B., Landström, H., & Persson, O. (2006). Entrepreneurial studies: The dynamic research front of a developing social science. *Entrepreneurship Theory and Practice, 30*(3), 375–398.

Dovi, E. (2006). Tapping women's entrepreneurship in Ghana: Access to credit, technology vital for breaking into manufacturing. *Africa Renewal, 20*(1), 12–15.

Dzanku, F. M., & Aidam, P. (2013). Agricultural sector development: Policies and options. In K. Ewusi (Ed.), *Policies and options for Ghana's economic development*. Accra: Institute of Statistical, Social and Economic Research.

Dzisi, S. (2008). *Women entrepreneurs in small and medium enterprises (SMEs) in Ghana*. Unpublished PhD thesis, Australian Graduate School of Entrepreneurship, Swinburne University of Technology, Australia.

Fosu, A., Aryeetey, E., & Quartey, P. (in press). Ghana's post-independence economic growth performance: A macroeconomic account. In G. Owusu, R. D. Osei, & F. A. Asante (Eds), *Contemporary development policies and practices: A reader*. Accra: University of Ghana.

Ghana National Development Planning Commission. (2010). *Medium-term national development policy framework: Ghana shared growth and development agenda (GSGDA), 2010–2013*. Accra: National Development Planning Commission.

Ghana Statistical Service. (2007). *Pattern and trends of poverty in Ghana, 1991–2006*. Accra: Ghana Statistical Service.

Ghana Statistical Service. (2014). *Ghana living standards survey round 6 (GLSS 6) – poverty profile in Ghana (2005–2013)*. Accra: Ghana Statistical Service.

Ghana Trades Union Congress. (2005). *Policies on employment, earnings and the petroleum sector*. Accra: Ghana Trades Union Congress.

Institute of Statistical, Social and Economic Research. (2010). *The state of the Ghanaian economy in 2009*. Accra: ISSER.

Institute of Statistical, Social and Economic Research. (2012). *The state of the Ghanaian economy in 2011*. Accra: ISSER.

Institute of Statistical, Social and Economic Research. (2014). *The state of the Ghanaian economy in 2013*. Accra: ISSER.

Krumina, M., & Paalzow, A. (2012). *Global entrepreneurship monitor: 2011 Latvia report*. Riga: TeliaSonera Institute.

Langevang, T. (2008). "We are managing!" Uncertain paths to respectable adulthoods in Accra, Ghana. *Geoforum, 39*(6), 2039–2047.

Langevang, T., & Gough, K. V. (2012). Diverging pathways: Young female employment and entrepreneurship in sub-Saharan Africa. *The Geographical Journal, 178*(3), 242–252.

Langevang, T., Gough, K. V., Yankson, P., Owusu, G., & Osei, G. (2015). Bounded entrepreneurial vitality: The mixed embeddedness of female entrepreneurship. *Economic Geography, 91*(4), 449–473.

Ministry of Youth and Sports. (2010). *National youth policy of Ghana: Towards an empowered youth, impacting positively on national development*. Accra: Ministry of Youth and Sports.

Ministry of Youth and Sports. (2013). *Ministerial impact assessment and review committee* on Ghana youth employment and entrepreneurial agency (GYEEDA)*. Retrieved 7 August 2015 from http://gbcghana.com/kitnes/data/2013/08/29/1.1505895.pdf.

Owusu, G., & Afutu-Kotey, R. L. (2014). Natural resources and domestic resource mobilization in Ghana: The case of the gold mining industry. In P. Quartey, E. B. D. Aryeetey, & C. G. Ackah (Eds), *Domestic resource mobilisation for inclusive development in Ghana* (pp. 338–367). Accra: Sub-Saharan Publishers/ISSER.

Owusu, G., & Lund, R. (2004). Markets and women's trade: Exploring their role in district development in Ghana. *Norwegian Journal of Geography, 58*(3), 113–124.

Owusu, G., & Oteng-Ababio, M. (2015). Moving unruly contemporary urbanism toward sustainable urban development in Ghana by 2030. *American Behavioral Scientist, 59*(3), 311–327.

Palmer, R. (2006). *Post-basic education, training and poverty reduction in Ghana* (PBET Policy Brief 3). London: DfID.

Robertson, C. (1995). Comparative advantage: Women in trade in Accra, Ghana, and Nairobi, Kenya. In B. House-Midamba & F. K. Ekechi (Eds), *African market women and economic power: The role of women in African economic development* (pp. 99–119). London: Greenwood Press.

Rosa, P., Balunywa, W., Dawa, S., Namatovu, R., Kyejjusa, S., & Ntamu, D. (2014). *Global Entrepreneurship Monitor 2012 report Uganda*. Kampala: Makerere University Business School.

Schoof, U. (2006). *Stimulating youth entrepreneurship: Barriers and incentives to enterprise start-ups by young people* (SEED Working Paper No. 76). Geneva: International Labour Office.

Xavier, S. R., Kelley, D. J., Herrington, M., & Vorderwülbecke, A. (2013). *Global Entrepreneurship Monitor: 2012 global report*. London: GERA.

Yankson, P. W. K., Owusu, G., Gough, K. V., Osei, R. D., & Langevang, T. (2011). *Global Entrepreneurship Monitor: GEM Ghana 2010 executive report*. Accra: ISSER.

Youth Enterprise Support. (2015). From ideas to reality: GH₵10 million fund to turn your idea into a business success. Retrieved 31 October 2014 from www.yes.gov.gh.

4 Measuring and promoting youth entrepreneurship in Zambia

Francis Chigunta and Valentine Mwanza

Introduction

Although Zambia has experienced rapid economic growth in recent years, the country continues to face the challenges of unemployment and poverty reduction. In particular, economic opportunities in the formal sector of the economy have been insufficient to absorb the large number of young people that either leave or drop out of the school system every year. This has given rise to concerns over the socioeconomic situation of the youth and the prospects of creating additional livelihood and employment opportunities for them. The difficult livelihood and employment situation among young people in Zambia appears even more discouraging when the fact that a large number of the youth are engaged in poor-quality jobs, with low salaries and no benefits or guarantees is taken into consideration. The government, unable to underwrite a labour market in which there is regular or secure employment in the formal sector, hopes that the youth unemployment problem will largely be solved through the promotion of youth entrepreneurship.

This chapter discusses the state of youth entrepreneurship in Zambia and presents an overview of key policies and interventions focusing on the promotion of youth employment and entrepreneurship. Drawing on the nationwide 2012 Global Entrepreneurship Monitor (GEM) survey, it scrutinises the levels and characteristics of youth entrepreneurship across the country, revealing the entrepreneurial attitudes of the youth, the nature of the businesses that they establish, the challenges and opportunities they face, and the measures being taken by the government and other stakeholders to promote youth entrepreneurship.

Socioeconomic context of youth entrepreneurship

Zambia's economy has experienced strong growth in recent years. The country's national output, as measured by gross domestic product (GDP), has grown by an average of 6–7% annually in the recent past. This has resulted in an economic performance that is higher than the regional growth levels for Southern Africa and sub-Saharan Africa of around 4% (Chigunta & Chisupa, 2013). The key contributors to the sustained increase in economic activity in Zambia have been increased production in the mining, agricultural, and construction sectors, with mining, fuelled by foreign investment, accounting for most of the growth. Prudent fiscal policy and sound macro-economic management have also been critical factors in creating an environment conducive to growth by stabilising economic conditions, lowering the user cost of capital and putting downward pressure on the real exchange rate. This has stimulated notable increases in foreign direct investment (FDI) flows (Chigunta & Chisupa, 2013). However, the most salient risks to Zambia's economy stem from global economic developments that directly impact the price of copper – Zambia's chief export. Any deterioration in the economic situation in the Eurozone, China, and other major markets can also adversely affect the economy.

The current pick-up in economic growth has propelled Zambia back into the ranks of middle-income countries. In the 1960s and early 1970s, Zambia was regarded as one of the richest countries in sub-Saharan Africa (Seidman, 1974). In 1975, the World Bank classified Zambia as a middle-income country with a per capita income of US$900. This period coincided with a favourable external environment characterised by high mineral rents. The government utilised the high mineral revenues to make heavy investment in social and economic infrastructure, especially in roads, schools, hospitals, electricity, and telecommunications. Although this investment was important, it bred aspects that have made the economy vulnerable to external shocks (Chiwele, 2000). The onset of the 1973 oil crisis, coupled with the collapse of the copper price, adversely affected the fiscal position of the Zambian government. Consequently, the economy went into secular decline in the 1980s and early 1990s, though this trend has been reversed in recent years.

A major challenge, however, is that Zambia's recent growth has not been inclusive. As used here, inclusive growth is taken to mean growth that not only creates new economic opportunities but also ensures equal access to these opportunities for all segments of society, especially the poor. While per capita income has risen from about US$370 in 2000 to over US$1,457 (2010 estimate) today, unemployment and poverty have stubbornly remained high. The 2010 Living Conditions and Monitoring Survey (LCMS) shows that poverty in Zambia is serious and widespread (CSO, 2012). While the poverty headcount declined from 68% in 2004 to 63% in 2010, it still remains high. According to the 2010 LCMS, extreme poverty stands at 43%. Notably, the statistics indicate that poverty is largely a rural phenomenon; while urban poverty declined from 53% in 2006 to 34% in 2010, there was an increase in rural poverty from 78% to 80% during the same period. Zambia, therefore, faces the challenge of ensuring that the benefits of growth are equitably shared by the entire population, including young people.

The labour absorptive capacity of the Zambian economy is limited. While the private sector is expected to create jobs, this has not been the case. The universe of Zambia's private sector is broadly divided into large enterprises and micro, small, and medium enterprises (MSMEs). The large enterprises generate most of the economic growth, exports, and tax revenues. However, they employ fewer workers than the small enterprises. The majority of MSMEs are found in the informal sector where most workers are unpaid family members or workers paid in kind who possess low levels of skills and education. The youth, especially those in the younger age cohorts, find themselves engaging in low-productivity, low-income survival pursuits in a range of activities, some licit and some illicit, in the informal economy (Chigunta, 2007; Gough, Chigunta & Langevang, 2015).

During the 1960s and 1970s, young people in Zambia did not pose a serious social problem. Since the mid-1970s, however, concerns have been rising over the socioeconomic situation of young people in Zanzibar and the prospects of creating additional livelihood opportunities for them (Chigunta, 2012; Chigunta, Chisupa & Elder, 2013; Hansen, 2010). The protracted and deep-rooted economic crisis that affected Zambia in the 1980s and 1990s, as noted above, adversely impacted on the well-being of the majority of people (Chigunta, 2007). As a consequence, many ordinary women and men have experienced a decline in their welfare owing to a fall in real incomes and declining social sector expenditure per head. However, it is youth, women, and other vulnerable people who seem to have particularly borne the brunt of the economic crisis and the measures adopted to restructure the economy, such as retrenchments and removal of subsidies (Argenti, 2002; Bennell, 2000; Kanyenze, Mhone, & Sparreboom, 1999). In the context of a high and growing incidence of poverty, and the documented adverse social impact of economic restructuring, there is increasing concern that large sections of young people have become marginalised and excluded from education, healthcare, salaried jobs, and even access to the social status of adulthood (Chigunta & Chisupa, 2013; Hansen, 2010; Locke & Lintelo, 2012). However, it is in the area of employment that young people have especially been affected.

Specific responses to the challenges facing youth vary from one African country to another but a notable convergence has been a growing interest in, and emphasis on, the role and potential of entrepreneurship in employment generation (Chigunta, 2007). The Zambian government hopes that its (youth) unemployment problem will be largely solved through the vigorous promotion of entrepreneurship, especially in the informal sector. In its National Youth Policy (NYP) adopted in 1994[1] and the National Programme of Action for Youth (NPAY) prepared in 1997, the government has explicitly stated that youth entrepreneurship should be actively promoted as a way of creating employment for young people. A crucial question that arises in this respect is the extent to which government policies and programmes have helped young people to go into entrepreneurship.

Characteristics of youth entrepreneurship in Zambia

After outlining the methodology of the GEM survey in Zambia, this section examines the entrepreneurial attitudes and perceptions of young people, levels of youth entrepreneurship, profiles of young entrepreneurs, and their growth aspirations.

Study methodology

The Global Entrepreneurship Monitor (GEM) is an annual household survey, which in 2012 was administered to 2,157 individuals (aged 18–64) in Zambia who were randomly selected and representative of the population. Using the standardised GEM Adult Population Survey (APS) questionnaire, data was collected through face-to-face interviews by trained field enumerators who translated the instrument into the local languages. Data was collected in all the country's nine provinces. The sampling was stratified at provincial, district, constituency, and ward level, using probability proportional to size. Three districts were selected in each province. A designated sample of households was taken within several selected enumeration units in each ward, and one person selected at random per household. The youth aged between 18 and 35 years accounted for 60% of the sample; the rest of the sample was made up of those aged 36–64 years.

Entrepreneurial attitudes and perceptions

The rates of perceived opportunities and capabilities among young people in Zambia are very high and broadly similar to those of the general population. According to the GEM survey, 84% of the youth felt that they had the necessary capabilities to start a business, while 79% claimed that there were good opportunities to start a business. This can, to an extent, be explained by the rapidly growing economy, coupled with the youth's general resourcefulness and openness to new ideas. However, more young men (53%) have better perceptions of opportunities than young women (24%). Similarly, more male youth (58%) believe that they have the necessary skills and knowledge to start a business than female youth (42%). Zambia's high rates of perceived opportunities and capabilities among young people suggest that the country has a large pool of potential young entrepreneurs. In such a context, people are likely to act on the opportunities they see, especially when they believe they are capable of starting a business. To a large extent, this may explain the high level of entrepreneurial activity in Zambia discussed in the next section.

The GEM data show that youth in Zambia have positive impressions about entrepreneurship as a good career choice (66% of all respondents). The youth also believe that entrepreneurs are afforded high status in society (78%) and receive positive media attention (70%). The level of fear of failure among young Zambians intending to start a business or running a business is among the lowest in the world (18% of all respondents). This further supports the view that youth in Zambia have positive impressions about entrepreneurship as a good career choice. Consequently, the proportion of youth with entrepreneurial intentions is high at 58%.

Level of entrepreneurship

The central measure of entrepreneurial activity used by GEM is the total entrepreneurial activity (TEA) rate, which measures the percentage of the adult population (18 to 64 years) who are in the process of starting or who have just started a business. This indicator, therefore, includes both "nascent" and "new entrepreneurs". Nascent entrepreneurs are those who have not paid salaries or wages for more than three months. New business owners are entrepreneurs who have moved beyond the nascent stage and have paid salaries and wages for more than three months but less than 42 months.

The 2012 GEM results show that, at 41%, Zambia has the highest early-stage entrepreneurship rate among GEM sampled countries in the world. At 42%, Zambia's TEA rate among young people aged between 18 and 35 years is even slightly higher than that of the general adult population. This means that approaching half of Zambia's youth population was involved in the process of starting or had recently started a new business during the previous year. The high TEA rate among young people in Zambia is reflected in the high incidence of start-up businesses (nascent entrepreneurs) and new firms (up to 3.5 years old) among the youth. The gender distribution of early-stage entrepreneurs among the youth shows that there were more males (59%) than females (41%) in 2012. The lower level of entrepreneurship among female youth could be explained by gender-related constraints arising from the different socialisation of young women and men where men are expected to be the breadwinners for the family.

The 2012 GEM data show that the number of established youth business owners (whose businesses are at least 3.5 years old) in Zambia is extremely low at 2%. This implies that, while Zambia has far more new young entrepreneurs, the overwhelming majority are failing to go beyond the early stages of enterprise formation. This supports previous research findings which suggest that the youth face far greater challenges and have far less support than their older counterparts (Chigunta, 2007).

Profiles of young entrepreneurs

The survey data show that almost half of young entrepreneurs (42.2%) have completed upper secondary school, 16.8% have attained a junior level of secondary education, and an insignificant proportion (0.4%) has never gone beyond primary school. A few young entrepreneurs (7.2%) have completed tertiary education, while 12.6% have had some form of tertiary education.

The majority of the young entrepreneurs (58.9%) in Zambia are involved in running businesses in the consumer-oriented sector (Table 4.1). According to the GEM classifications, this sector comprises retail, motor vehicles, lodging, restaurants, personal services, health, education and social services, and recreational services. Young people are more likely to find themselves in this sector, especially retail or trading, due to ease of entry (Chigunta, 2007). However, the disaggregation of the data show that there are significantly more female youth

Table 4.1 Sector distribution of youth TEA entrepreneurs in Zambia (%)

Sector	Youth TEA entrepreneurs (aged 18–35)
Extractive (forestry, agriculture, fishing, mining)	15.4
Transforming (construction, manufacturing, transportation, communication, utilities, wholesale)	22.2
Business services (finance, insurance, real estate, all business services)	3.6
Consumer-oriented (retail, motor vehicles, lodging, restaurants, personal services, health, education and social services, recreational services)	58.9

Source: Zambia GEM survey (2012).

(65.2%) than male youth (55.1%) operating businesses in this sector. In contrast, more young men (24.2%) than young women (18.5%) operate businesses in the transforming sector comprising construction, manufacturing, transportation, communication, utilities, and wholesale. These gender differences suggest that female youth tend to face more difficulties than male youth in getting into other business sectors apart from trading. Other sectors are the extractive sector (i.e. forestry, agriculture, fishing, and mining) and business services (i.e. finance, insurance, real estate, and all business services).

It is, however, encouraging to note that there are more young entrepreneurs who see themselves as pursuing opportunities than those who are pushed by necessity. In the GEM survey, nearly half of the early-stage young entrepreneurs in Zambia (45%) reported that their involvement in entrepreneurship was opportunity-driven while about one-third (32%) indicated that they were driven into entrepreneurship out of necessity. The great majority of the opportunity-driven young entrepreneurs were pulled by an opportunity to increase income (74%), followed by the desire for increased independence a distant second place (20%). The remaining few (5%) were motivated by the need to maintain their income; that is, to ensure that their income does not fall below a certain level if it does not increase. The influence of friends and family was paramount in relation to young people establishing a business. Half (50%) of the young people stated that they were primarily influenced by friends to start their business and 46% were influenced by a range of family members including siblings, parents, grandparents, and other relatives.

The GEM data show that the majority (69%) of the youth used personal savings to start their business. The next important source was support from family members (18%) with a further 2% turning to friends for financial support. Only 11% of the young entrepreneurs accessed funding from a bank or other financial institution. There are no significant variations in identified sources of income between the youth who are already involved in business and those who are planning to establish a business. This indicates that young people in Zambia are aware of the difficulties that they face in accessing formal bank capital to start their business, hence they find alternative sources. It is worth noting that only a tiny proportion of the young entrepreneurs (5%) stated that their primary sales come from family and friends, and similarly only 2% of young people intending to go into business indicated that their sales would be likely to come from family or friends. This is an indication of need for young people to find consumers beyond their immediate circle of family and friends when establishing a business.

The majority of young entrepreneurs can only create employment for themselves. Their growth expectations are also limited, with less than one-third (32%) indicating they expect to add no more than five employees within the next five years and less than 1% (0.5%) project having 20 or more employees. Despite this, the majority of young entrepreneurs (67%) stated that they would prefer to run their own business as a long-term option. Less than one-quarter (22%) expressed interest in being employed by others, while a small proportion (11%) was undecided. In gender terms, more male youth (57%) indicated that they would prefer to run their own business as a long-term option than female youth (43%). An even higher proportion of the youth not involved in entrepreneurship (77%) indicated that they would prefer to run their own business as a long-term option as opposed to being employed by others (21%), while a tiny proportion (2%) was undecided. It is, therefore, important to tap into the entrepreneurial intentions of the youth and provide them with optimum chances of growing and running viable businesses within a supportive environment. If this is not done, chances are high that they could become disillusioned and negative, with some choosing to discontinue their business.

Youth entrepreneurship policies and support programmes

The growing problem of youth unemployment in Zambia has started to receive serious attention at the highest government level. Both the state and non-state actors, among them nongovernmental organisations (NGOs), have initiated a number of programmes designed to promote youth employment through entrepreneurship promotion. Despite these initiatives, there is currently an incoherent approach towards addressing the youth employment challenge in Zambia at the institutional level. While the relevant Policy, Legal and Regulatory Framework (PLRF) for the promotion of youth employment creation and entrepreneurship development exists, it lacks the synergies necessary to facilitate concerted action. Moreover, young people are not being actively engaged as partners in the process of policy and programme implementation hence are unable to assert their role in informing the development of relevant strategies to address their concerns.

The youth specific policy framework in Zambia comprises the NYP and the NPAY. However, the NYP has not been effective in serving as a guiding framework for youth policing and programming due to a number of factors, including lack of adequate funding. According to Chigunta and Chisupa (2013), the implementation of the NYP is characterised by a low level of youth participation, absence of periodic reviews, lack of policy dialogues involving Ministry of

Youth and Sports officials and major stakeholders, lack of linkages with sectoral policies, and absence of strategic coordination. The situation of youth has not been helped by the problematic implementation of the NPAY. Launched in 2007, the NPAY was envisioned as a tool to operationalise the NYP into what it calls "actionable" units. Besides funding, there were other critical challenges hampering the effective implementation of the NPAY, many of which emanate from weaknesses in the NYP (Figure 4.1).

A key challenge is the education and training system. Despite recent progress, the education sector in Zambia is characterised by inadequate education inputs leading to poor quality of education outcomes, lack of relevance of education outcomes to labour market needs, limited linkages with industry and the world of work, and lack of entrepreneurship training and career guidance. In addition, while Zambia may have what looks like a very elaborate technical education and vocational skills training (TEVET) system under the Technical Education, Vocational and Entrepreneurship Training Authority (TEVETA) in the Ministry of Education, the country is currently facing the twin crises of a high youth unemployment rate and a shortage of job seekers with employable skills. The TEVET sector is beset with many challenges including: inadequate funding, poor service providers, inadequate information management system, inadequate staff, limited training equipment and materials, access and equity problems, concerns over the quality of graduates, and underdeveloped links with industry. Moreover, there is presently no clearly defined human resource development strategy that allies the education and training offered to young people with the objectives of the country's national development plans (Chigunta & Chisupa, 2013).

The government has, however, initiated enterprise development policies and programmes aimed at promoting employment creation. Key policies and legislation in

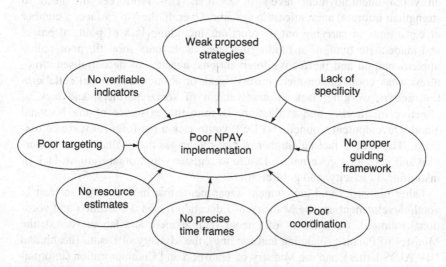

Figure 4.1 Factors negatively affecting the implementation of the National Plan of Action for Youth (source: Chigunta & Chisupa, 2013).

this regard include the Micro, Small and Medium Enterprises (MSME) Policy, the Citizens Economic Empowerment Commission (CEEC) Act and the Zambia Development Agency (ZDA) Act. Although data is not available on the impact of the enterprise development policies, it is evident that small enterprises face many challenges, and employment creation through enterprise formation in the formal sector is very limited. While various policies have been developed, the government lacks the ability to effectively implement them.

Since the mid-1970s, Zambia has also operated a series of work experience and training programmes without much success. Some forms of active labour market policies (ALMPs) exist but evaluations of these policies are generally non-existent. Thus, there is little clarity on which practices and interventions work and which can be scaled up. Most skills initiatives in Zambia today serve a few hundred or perhaps a few thousand youth. The absence of a clear policy framework on ALMPs, such as apprenticeships, is hindering the expansion of these measures to benefit more young people.

Presently in Zambia, there is no clearly identified agency that is responsible for coordinating issues related to youth employment and entrepreneurship promotion. As in the case of policy, the institutional framework for youth employment in Zambia can be divided into two: youth-specific and general support institutions. The former largely consists of the Ministry of Youth and Sports (MoYS). According to the NYP, among others, the MoYS is responsible for youth policy coordination and runs a network of 17 youth resource centres across the country that provide out-of-school youth with training in life skills, basic technical skills, and entrepreneurship. The annual intake of these centres is about 1,000 students, which, in view of the over 300,000 young people who leave the school system every year, is too low to have any significant impact on youth unemployment levels in Zambia. This reinforces the need to strengthen informal apprenticeship systems. Overall, the MoYS faces a number of constraints in carrying out its functions including: lack of political power and inadequate funding; a funding regime that favours football; poor policy implementation and ineffective interventions; absence of decentralised structures; weak coordination mechanisms; weak or absent institutional collaboration and networking; lack of involvement of social partners; and lack of effective monitoring and evaluation. Despite the existence of the National Youth Development Council (NYDC), youth lack a platform for national dialogue. The NYDC faces a number of key challenges that include weak leadership and corporate governance, failure to register youth organisations, lack of institutional capacity, and politicisation.

Other ministries and government departments that have a major interest in youth development are the Ministry of Education (general education and vocational training), the Ministry of Labour (employment and labour issues), the Ministry of Finance (planning and funding), the Ministry of Health (health and HIV/AIDS issues), and the Ministry of Transport and Communication (information and communication technology). The work on youth by these ministries is largely fragmented and, as previously noted, in Zambia there is presently no clearly identified agency that is responsible for coordinating issues related to youth employment and entrepreneurship promotion. Moreover, there is poor representation of what the International Labour Organization calls social partners in the formulation and implementation of youth policies and programmes on employment creation and entrepreneurship promotion (Chigunta & Chisupa, 2013).

Conclusions

After decades of economic decline, in recent years Zambia's economy has been experiencing strong growth mostly driven by the high prices of the country's main export, copper. However, the country's growth has not been inclusive and unemployment and poverty have remained high. This situation has compelled unemployed young people, especially those not at school, to venture into self-employment and entrepreneurship in a range of activities, some licit and some illicit, in what is generally referred to as the informal sector.

The 2012 GEM survey findings show that the youth have a very high perception of the presence of good opportunities for starting businesses. The overwhelming majority are also likely to believe that they have the capabilities necessary to start a business. Equally, entrepreneurial intentions – that is, the desire to start a business – are high and few young Zambians fear business failure. Societal impressions about the attractiveness of entrepreneurship in the country remain strong, with the majority of young Zambians seeing entrepreneurship as a good career choice. Not surprisingly, there is a high level of dynamic entrepreneurial activity among young people aged between 18 and 35 reflected in the high incidence of start-up businesses (nascent entrepreneurs) and new firms (up to 3.5 years old). However, the rate of business discontinuance is among the highest in the world, resulting in the established business rate among young entrepreneurs in Zambia being among the lowest. This can be attributed to the absence of a strong supportive environment. Although there are some institutions, including governmental and NGOs that seek to address youth employment in Zambia, they are few and far between. Moreover, the uptake and coverage of existing youth enterprise development programmes is very limited, hence their impact appears minimal.

In view of the above, a number of recommendations emerge. There is need for a deliberate media campaign aimed at raising awareness about existing youth entrepreneurship support programmes among young people. It is important that these entrepreneurship support programmes are not overly bureaucratic and are easily accessible to young entrepreneurs. In particular, the government and other organisations involved in supporting young entrepreneurs should expand their interventions aimed at assisting young people intending to go into entrepreneurship by providing access to capital. There is also a need to encourage the establishment of a wide-ranging apprenticeship system to provide young entrepreneurs with appropriate skills. Furthermore, the existence of role models who would remain constantly in touch with young people in order to give them business guidance and encouragement would be invaluable.

Note

1 In 2004, the NYP was revised to take into account changes that have occurred over the past 10 years.

References

Argenti, N. (2002). Youth in Africa: A major resource for change. In A. de Waal and N. Argenti (Eds), *Young Africa: Realising the rights of children and youth* (pp. 123–153). Trenton: Africa World Press.

Bennell, P. (2000). *Improving youth livelihoods in SSA: A review of policies and programmes with particular emphasis on the link between sexual behaviour and economic well-being.* Report to the International Development Center (IDRC).

Central Statistics Office (CSO). (2012). *Living conditions monitoring survey VI 2010.* Lusaka: Central Statistics Office.

Chigunta, F. (2007). *An investigation into youth livelihoods and entrepreneurship in the urban informal sector in Zambia.* Unpublished doctoral thesis, Oxford University.

Chigunta, F. (2012). *The youth and labour market needs in Zambia.* Lusaka: British Council.

Chigunta, F., & Chisupa, N. (2013). *Review of the effectiveness of youth employment policies, programmes, strategies and regulatory framework.* Lusaka: International Labour Organization.

Chigunta, F., Chisupa, N., & Elder, S. (2013). *Labour market transitions of young women and men in Zambia* (Work4Youth Publication Series No. 5). Geneva: International Labour Organization.

Chiwele, D. K. (2000). *Investment for poverty reducing employment: Industrial investments in Zambia.* Report to International Labour Organization, Geneva.

Gough, K. V. (2008). "Moving around": The social and spatial mobility of youth in Lusaka. *Geografiska Annaler: Series B, Human Geography, 90*(3), 243–255.

Gough, K. V., Chigunta, F., & Langevang, T. (2015). Expanding the scales and domains of insecurity: Youth employment in urban Zambia. *Environment and Planning A*, DOI: 10.1177/0308518X15613793.

Hansen, K. T. (2010). Changing youth dynamics in Lusaka's informal economy in the context of economic liberalisation. *African Studies Quarterly, 11*(2–3), 13–27.

Kanyenze, G., Mhone, G. C. Z., & Sparreboom, T. (1999). *Strategies to combat youth unemployment and marginalisation in anglophone Africa.* Geneva: ILO/SAMAT.

Locke, C., & Lintelo, D. J. (2012). Young Zambians "waiting" for opportunities and "working towards" living well: Lifecourse and aspiration in youth transitions. *Journal of International Development, 24*(6), 777–794.

Seidman, A. (1974). The distorted growth of import-substitution industry: The Zambian case. *Journal of Modern African Studies, 12*, 601–631.

Concluding comments to Part I

Katherine V. Gough and Thilde Langevang

Despite different levels of development in the three countries, with Ghana and Zambia being classified as lower middle-income countries while Uganda is classified as a low-income country, as the three preceding chapters have shown, the levels and characteristics of youth entrepreneurship and attitudes to starting a business among young people are remarkably similar. Some subtle differences, however, have emerged that stem from each country's particular political, economic, and cultural histories. In this concluding commentary to Part I, we draw out these comparisons and place the observed trends in entrepreneurship levels and policies in a global perspective.

With TEA entrepreneurship levels ranging between 39 and 42%, the youth of Uganda, Ghana, and Zambia exhibit similar high levels of entrepreneurship. These levels are equal to or higher than the adult population in their respective countries, and much higher than their youth counterparts in other regions of the world where average rates range from just 9% in the Middle East and North Africa to 18% in Latin America (see Figure PI.2). In comparison with the other sub-Saharan African countries participating in GEM, Zambian, Ghanaian, and Ugandan youth scored the highest, second highest, and third highest rates, respectively (Kew, Herrington, Litovsky, & Gale, 2013).

Young people in Uganda, Ghana, and Zambia generally have very positive attitudes towards self-employment; they have a relatively high confidence in their ability to start a business and they perceive the opportunities for entrepreneurship to be favourable. For the majority of young Ugandans, Ghanaians, and Zambians, starting a business is seen not only as a legitimate livelihood option that many are willing to pursue but it is actually their preferred employment avenue. Compared to other sub-Saharan African countries, the attitudes towards entrepreneurship of young people in all three countries studied are slightly higher while they are significantly higher than the averages for other regions of the world (Kew et al., 2013). Despite their enthusiasm for starting a business, however, a common characteristic of young entrepreneurs in all three countries is that their businesses stay at the micro-level. This is particularly due to their businesses being concentrated in trading where competition is high and earnings minimal. The majority of young entrepreneurs have no or only a small number of employees hence contribute little to job creation apart from

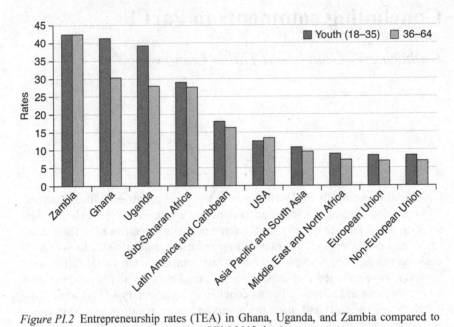

Figure PI.2 Entrepreneurship rates (TEA) in Ghana, Uganda, and Zambia compared to regional averages (source: GEM 2012 data).

self-employment, have low expectations for growth, and their businesses close down at a high rate.

One noticeable difference between the three countries is Ghana's higher rate of female entrepreneurship. While all three countries have rates of female entrepreneurship that are much higher than in most other parts of the world, in Ghana exceptionally the rate of female youth entrepreneurship exceeds that of males. This is historically rooted with women having been very active in trading since pre-colonial times. This construction of female identity as being compatible with business ownership has deepened in Ghana in recent decades linked to structural adjustment programmes and economic restructuring, which have resulted in more women having to work to support the household (Langevang, Gough, Yankson, Owusu, & Osei, 2015).

The chapters also revealed how all three countries have witnessed a similar mushrooming of entrepreneurship promotion initiatives in recent years instigated by governments, NGOs, private sector organisations and multilateral organisations. While the availability, aim, and scale of their schemes vary, the general picture emerging is that youth entrepreneurship promotion is characterised by a plethora of largely uncoordinated schemes, which tend to have limited uptake and scope, some of which have become highly controversial in the eyes of the public. As Honwana (2012) similarly finds, youth entrepreneurship schemes in Africa are often politicised and used as instruments for political control and clientelism, and can become objects for corruption and mismanagement of funds.

There is clearly a dire need to de-politicise youth entrepreneurship programmes in sub-Saharan Africa to ensure that beneficiaries are not being used to promote political agendas and that it is not only a select group of well-connected young people who benefit.

Examples of policies and support measures focusing on the five crucial factors for entrepreneurship engagement as identified by Schoof (2006) – social and cultural attitudes towards youth entrepreneurship, entrepreneurship education, access to financing, administrative and regulatory frameworks, and business assistance and support – exist in all three countries. Since attitudes to starting a business of one's own are generally very positive across the three countries studied, and entrepreneurship rates among the youth are generally very high, the focus of policy makers should be expanded to include a range of aspects instead of focusing narrowly on promoting start-ups. Business survival and growth in particular appear as key challenges for youth-run businesses across the three countries, issues that need to be addressed. The rationale for designing particular interventions focusing on young people is that they have less experience, limited capital, fewer social network relations, and a lower social position in society. However, as Izzi (2013) also highlights, it is important to acknowledge that young people do not form a homogeneous group of entrepreneurs and that the incentives and barriers they experience may be differentiated by, for example, gender, place, age, and sector. Consequently, a more holistic approach to entrepreneurship promotion is needed, which recognises the multiple factors that impact on young people's ability to start and grow viable businesses, and which are sensitive to the varying needs of different segments of the population.

The problem of having numerous isolated, small-scale initiatives and pilot programmes promoting youth entrepreneurship is not unique to Uganda, Ghana, and Zambia but rather seems to be a recurrent issue worldwide. In a global mapping of good examples, practices, and initiatives for promoting youth entrepreneurship, Schoof (2006: 73) finds "incoherency, poor co-ordination and the lack of synergy" between projects to be a major problem. Given the multitude of uncoordinated initiatives there is a need to consider adopting horizontal policy making and coordination mechanisms. At the governmental level this implies collaboration between different ministries, such as education, finance, labour, youth, and industry, and filling in the gaps between high-level national policies for enterprises/entrepreneurship and initiatives operated by other stakeholders to ensure coordination and coherence. Since youth entrepreneurship policies are inherently cross-cutting, the numerous governmental and nongovernmental stakeholders need to adopt a collaborative approach.

The problem of ineffective policies is compounded by the fact that there are strikingly few evaluations of the impact of such schemes in reaching their aim of resolving the youth unemployment situation through entrepreneurship promotion. Assessments and evaluations of programmes are needed both to improve entrepreneurial conditions and to ensure the cost-effective use of funds. Evaluating the impact and effectiveness of initiatives to promote entrepreneurship is not an easy task and requires methodological innovation (Schoof, 2006). Moreover, it is vital that entrepreneurship is not perceived as the single solution to the youth unemployment problem but rather is seen as just one element of broader labour market policies, which cannot be abstracted from wider policies aimed at stimulating job-generating, inclusive, and sustainable economic growth (Chigunta, Schnurr, James-Wilson, & Torres, 2005; Gough, Langevang, & Owusu, 2013). Without this shift in approach, policies to promote youth entrepreneurship will continue to miss their target and young entrepreneurs will go on struggling without the assistance they need.

References

Chigunta, F., Schnurr, J., James-Wilson, D., & Torres, V. (2005). Being "real" about youth entrepreneurship in eastern and southern Africa (SEED Working Paper 72).

Gough, K. V., Langevang, T., & Owusu, G. (2013). Youth employment in a globalising world. *International Development Planning Review, 35*(2), 91–102.

Honwana, A. M. (2012). *The time of youth: Work, social change, and politics in Africa.* Sterling, VA: Kumarian Press.

Izzi, V. (2013). Just keeping them busy? Youth employment projects as a peacebuilding tool. *International Development Planning Review, 35*(2), 103–117.

Kew, J., Herrington, M., Litovsky, Y., & Gale, H. (2013). *Generation entrepreneur? The state of global youth entrepreneurship* (The Prince's Youth International Business (YBI) and Global Entrepreneurship Monitor (GEM) Report).

Langevang, T., Gough, K. V., Yankson, P. W., Owusu, G., & Osei, R. (2015). Bounded entrepreneurial vitality: The mixed embeddedness of female entrepreneurship. *Economic Geography, 9*(4): 449–473.

Schoof, U. (2006). *Stimulating youth entrepreneurship: Barriers and incentives to enterprise start-ups by young people* (No. 388157). Geneva: International Labour Organization.

Part II

Youth entrepreneurship in urban settlements

Introduction to Part II

Katherine V. Gough and Thilde Langevang

Despite high rates of urban growth in sub-Saharan Africa being widely proclaimed, with accompanying scenarios of cities growing out of control with ever-expanding slums, these growth rates are contested amid claims that they are often based on the misinterpretation of unreliable data (Satterthwaite, 2007). Potts (2009, 2013) in particular has conducted detailed work that illustrates how sub-Saharan African urbanisation rates are highly variable. Recent censuses show that many urban growth rates are around or below the national growth rates with much urban growth today being due to natural population increase rather than migration from rural to urban areas. While Ghana is classified among sub-Saharan African countries that have experienced recent rapid urbanisation, Uganda falls into the category negligible urbanisation, while Zambia has experienced periods of counter-urbanisation (Potts, 2013). Across sub-Saharan Africa, however, cities are predominantly and increasingly inhabited by young people; while the population in most of the rest of the world is ageing, half the population in sub-Saharan Africa is under 25 years of age (Filmer & Fox, 2014), many of whom are born in or move to urban areas at some point in their lives.

In urban areas, the implementation of structural adjustment policies (SAPs) in the 1980s and 1990s removed a key employer of educated workers and left a large number of public sector workers and civil servants jobless, trends that have continued under policies of neoliberalism. Consequently, informality has increased and the informal economy is currently the main source of employment. Due to its very nature, determining the proportion of the population working in the informal economy is challenging; however, it is estimated that as many as nine in 10 urban workers in Africa have informal jobs (ILO, 2009). Although informality has now come to be accepted as the norm in African cities, the inadequacy of the terms "formal" and "informal" to describe the nature of urban employment needs to be noted. Not only do people combine work in formal and informal economies, the boundary between the two is indistinct and overlapping with many so-called informal enterprises being registered and taxed in one form or another (Lindell, 2010). The vast majority of young people entering the workforce in urban areas have little choice other than to work in the informal economy.

While informality has tended to be perceived as negative, with the ideal being formal public or private sector employment, increasingly self-employed individuals are being viewed as entrepreneurs. As the chapters in Part I showed, entrepreneurship is being promoted as the way forward to both stimulate economic growth and generate youth employment, especially in urban areas. In Part II we turn to focus on young entrepreneurs in urban areas, exploring why and how they establish businesses and the challenges they face. In each country, one low-income settlement in the capital city is the focus of the research, where both quantitative and qualitative. data were collected working in collaboration with youth peer researchers. The use of peer researchers has recently gained popularity, especially in research involving marginalised and vulnerable groups such as children and youth, as it enables individuals from the research target group to work alongside professional researchers in exploring issues central to their lives (Robson, Porter, Hampshire, & Bourdillon, 2009). The peer researchers were involved in several stages of the research process including drawing up the research instruments and carrying out field research. The chapters show how, despite facing many hurdles, young people are highly resourceful in perceiving new opportunities, altering or switching their businesses as circumstances change.

References

Filmer, D., & Fox, L. (2014). *Youth employment in Sub-Saharan Africa.* Washington DC: IBRD/World Bank.

ILO. (2009). *The informal economy in Africa: Promoting transition to formality – Challenges and strategies.* Geneva: International Labour Organization.

Lindell, I. (2010). Changing landscapes of collective organizing in the informal economy. In I. Lindell (Ed.), *Africa's informal workers: Collective agency, alliances, and transnational organizing* (pp. 4–34). Uppsala: Nordic Africa Institute Press.

Potts, D. (2009). The slowing of sub-Saharan Africa's urbanization: Evidence and implications for urban livelihoods, *Environment and Urbanization, 21*(1), 253–259.

Potts, D. (2013). *Rural–urban and urban–rural migration flows as indicators of economic opportunity in sub-Saharan Africa: What do the data tell us?* (Working Paper 9, Migrating out of Poverty). University of Sussex.

Robson, E., Porter, G., Hampshire, K., & Bourdillon, M. (2009). "Doing it right?" Working with young researchers in Malawi to investigate children, transport and mobility. *Children's Geographies, 7*(4), 467–480.

Satterthwaite, D. (2007). *The transition to a predominantly urban world and its underpinnings* (Human Settlements Discussion Paper Series, Theme Urban Change), 4.

5 Young entrepreneurs in Lusaka

Overcoming constraints through ingenuity and social entrepreneurship

Francis Chigunta, Katherine V. Gough, and Thilde Langevang

Introduction

Young people in Zambia are growing up in an environment where their job prospects are worse than those of their parents. The fall in copper prices in the 1970s and the introduction of structural adjustment programmes in the 1980s resulted in a severe decline in the number of both public and private sector jobs (Maipose, 1989). Young people's education is still geared towards obtaining formal employment but given that the majority of the labour force is informally employed, only a minority of youth stand the chance of gaining formal employment. The vast majority of young people living in urban settlements work for informal enterprises (which are unregulated, with very low, irregular pay), manage to establish a business of their own, or are unemployed. Although the Zambian economy has recovered somewhat in recent years, with GDP growth rates of 6–7% enabling the country to rejoin the ranks of middle-income countries, this has not translated into any substantial growth in the number of jobs. Moreover, few families in Zambia remain unaffected by the HIV/AIDS pandemic, which has resulted in many young people growing up without one or both parents and having to bear the associated social and economic consequences (Day & Evans, 2015).

Most residents of Lusaka, the capital city of Zambia, live in informal settlements (referred to as compounds) characterised by poor quality, overcrowded housing, and inadequate services (Schlyter, 2011). This chapter draws on research conducted in one such compound, Chawama, in order to explore the employment situation of young people. While recognising that life is extremely tough for many youth, we show how young people who live in a challenging environment and face many hurdles can be resourceful, turning constraints into opportunities. We also highlight how some young people in the community act as "social entrepreneurs" – a term used to denote entrepreneurial activity driven by a mission to create social value and social change, rather than only generate economic wealth for the business owner (Mair and Marti, 2006). Through their actions, the young entrepreneurs seek not only to improve their own situation, but also that of others less fortunate than themselves.

Chawama compound

Located in a productive farm area in central Zambia, Lusaka is an administrative, financial, and commercial centre. The city, which covers an area of 360 km^2, lies at the junction of the Great North Road (to Tanzania) and the Great East Road (to Malawi) and is on Zambia's main railroad. As in the case of other major cities and towns in southern Africa, Lusaka was founded by the Europeans as a small railway outpost in 1905. In the 1930s, Lusaka was planned according to the ideals of a British garden city with open spaces, large building plots for Europeans, and limited planned housing on much smaller plots for Africans. Squatter settlements grew up on the outskirts of the city to house the many Zambians who had moved to Lusaka in search of work (Hansen, 1997). These settlements, which have grown in number over time, form Lusaka's current 37 compounds. Today, about 40% of Zambia's population lives in urban areas; Lusaka is by far the largest urban centre, with a population of almost 2.2 million. Approximately 80% of its population live in poor, unplanned settlements (also known as *ma komboni*), which constitute about 20% of the city's residential land (Central Statistical Office, 2012).

The data for this study was collected in Chawama, a large compound that covers an area of 8 km^2 on the outskirts of the city. There are few industries or formal businesses located in Chawama, hence residents depend on the informal economy and the wider urban setting for their livelihoods. Census data indicate that the population of Chawama rose from a mere 5,500 in 1965 to 23,000 in 1974, and from around 60,000 in 1990 to an estimated almost 100,000 today. The quality of service provision in Chawama remains poor despite a World Bank upgrading project in the 1970s (Moser & Holland, 1997). During the rainy season, the area is severely flooded, impeding movement throughout the settlement and spreading waterborne diseases. Although most residents can be classified as low income, some middle-income households have moved into the compound due to their worsening financial situation and the overall housing shortage in Lusaka. Chawama compound thus contains residents of diverse socioeconomic, ethnic, religious, and regional backgrounds.

In line with the Youth and Employment (YEMP) project, this chapter adopts a broad and inclusive conceptualisation of entrepreneurship as being practices that include some form of business or trading activity engaged in young people to make a living. These practices include setting up an enterprise structure (such as a hair salon) or trading goods, as well as more ephemeral or dubious forms of business activity. Social entrepreneurship is the term given to entrepreneurs who have the capacity to recognise and take advantage of opportunities to create social change. While social entrepreneurs are found in a range of settings, they are typically associated with contexts such as Chawama, which are characterised by scarce resources and emerge in response to deficient facilities and services in such communities (Di Domenico, Haugh, & Tracey, 2010; Mair & Marti, 2006).

Methodology

Chawama compound was selected for this study because the first author has conducted research in the area since 2000, and because it demonstrates many characteristics typical of informal settlements in Lusaka. Quantitative and qualitative data were collected in two separate phases by the authors, working in close collaboration with young people living in the compound who were employed as research assistants. As other studies of young people in sub-Saharan Africa show, this form of participatory, peer-led research not only tends to result in a high response rate, but can also generate the best insights (Gough, Langevang, & Namatovu, 2014; Porter & Abane, 2008; Robson, Porter, Hampshire, & Bourdillon, 2009). The first author was responsible for developing and carrying out a questionnaire survey in October 2010. The aim of the survey was to find out which activities young people engaged in, the constraints they faced in setting up and running a business, and their use of support mechanisms. The respondents were randomly selected young people living in Chawama, aged 15–35 in line with the Zambian definition of youth. A total of 369 questionnaires were administered by 19 peer researchers from Chawama once they had been appropriately trained. Each questionnaire typically took 30–40 minutes to complete, mostly in respondents' homes.

The qualitative data were collected during the second phase of fieldwork in October 2012. Following introductions to key actors in Chawama by the first author, the second and third authors conducted life-story interviews and focus group discussions (FGDs) with assistance from the peer researchers, who acted as interpreters where necessary. A total of 20 young people running different types of businesses were selected for in-depth interviews, taking care to cover both genders and a range of ages within the youth category. Adopting a life-story approach, they were encouraged to recount in detail a range of aspects of their lives, including changes in their family and housing situation over their lifetime, their education and work experiences. Respondents were interviewed either at home or at work. Around half the interviews were taped and subsequently transcribed; due to the challenges of recording in noisy settings, detailed notes were taken during the other interviews. A total of five FGDs were held with the aim of hearing young and older Chawama residents talk about the situation of youth employment. The three FGDs with youth involved mixed-gender groups of around six to nine young people as they had expressed a preference for both young men and women to be present. The two FGDs with older people were single-sex groups of four to seven people to ensure that women, who tend to be more reticent when men are present, had a chance to have their say. The FGDs were held on the premises of the Chawama Youth Project (see below) and lasted about 1 hour each. In line with common practice, participants were offered refreshments. The discussions were taped and subsequently transcribed.

Entrepreneurial activities of Chawama youth

When asked what their main livelihood was, one-quarter of young people in Chawama were found to be running an enterprise (Table 5.1). A further 8% were employed either formally or informally and, similarly, 8% were engaged in piecework – which, in the Zambian context, is casual work that carries a daily wage (known locally as *kubazabaza*). Almost one-third claimed to be "doing nothing". When combined with those who said that housework was their main activity, almost half the young people were either engaged in unpaid housework or appeared to be idle. It is important to recognise, however, that "doing nothing" – sometimes referred to as "sitting at home" – in Zambian parlance rarely means complete idleness. Rather, it covers highly irregular casual work, very small and ephemeral enterprise activity, and volunteer work (Gough, 2008; Hansen, 2005). In this chapter, however, the primary focus is on young people who are enterprise owners.

The young people themselves had quite a nuanced view of what they considered an entrepreneur to be. Various perceptions of entrepreneurs emerged during the FGDs but young respondents primarily saw an entrepreneur as somebody who starts a business and makes a profit:

> I think an entrepreneur is a person who has got an idea or ideas about how to start a business.
>
> (Female, 23)

> I'll add on, it's somebody who has an idea, creative, innovative ideas how to run a business with the purpose of making profit. Because if you are into business but you are not making a profit, then you cannot qualify to be a businessman. If you are doing business and you are making a profit to sustain your daily living, then you are an entrepreneur.
>
> (Male, 29)

Slightly more young men (28%) reported running a business than young women (22%), which is in line with the Global Entrepreneurship Monitor data for

Table 5.1 Employment status of young people in Chawama

Employment status	Percentage
Enterprise owners	25
Employed	8
Piece work	8
Students	18
Housework	17
"Doing nothing"	32
Sample size	350

Source: YEMP questionnaire survey (2010).

Zambia (see Chapter 4). Most business owners (84%) were engaged in small-scale trading, followed by services (13%), while just 3% were involved in manufacturing. The most common trading activities were: selling fresh vegetables, fruit, fish, or meat; preparing and selling food and snacks; providing daily groceries such as oil, salt, and soap; trading *salaula* (second-hand clothes); brewing and selling *kachasu* (liquor); and trading electronics, including mobile phones, accessories, and airtime (pre-paid phone cards). Common services provided by entrepreneurs included hairdressing, barbering, shoe mending, car washing, education/training services, and restaurants and bars, while manufacturing included carpentry and welding.

Around one-third of the young people's enterprises were based in a marketplace (32%), with slightly fewer operating from the roadside (29%) or from home (25%). A further 8% were mobile with no fixed location. Most activities took place within the local community with few entrepreneurs having any direct links to the formal sector; less than one in 10 had received orders or contracts to supply products or services to a private formal company and only two in 10 had sourced items or raw materials from local or foreign private formal enterprises. Most relied on locally, informally sourced goods.

Just over half the young people had acquired their start-up capital from their own savings (51%) and a further one-fifth had received financial support from parents or relatives (21%). Those who had used their savings had raised the money primarily through previous employment either in informal jobs (23%), through piecework/casual labour (30%), or through a formal wage job (16%). This shows how many young people work for some time before setting up a business and use their accumulated savings or rely on family support structures for start-up capital. The initial investment put into these businesses was generally very low: two-thirds (65%) had started with less than ZMK500,000 and 15% had set up with only ZMK50,000 or less.[1] None of the young people had obtained a loan from a formal institution. As a young man explained, formal financial institutions tended to be interested only in supporting those who were already running a business:

> Most of the time they look at somebody who is already doing a business.... But if I don't have anything, it will be too difficult for me to access those funds because they'll want somebody who is already into business.
>
> (Male, 29)

Most of these enterprises had been launched relatively recently; just over half (52%) had operated for less than three years while one-fifth (21%) had existed for a year or less. Their earnings were relatively low with more than half (57%) making profits of less than ZMK300,000 a month. Moreover, the enterprises were generally very small and generated limited additional employment: almost 80% were run by the entrepreneur alone or relied on unpaid family labour, an additional 7% had unpaid apprentices, 2% occasionally employed casual workers, and just 3% had any permanent employees. The overwhelming majority

(93%) of young entrepreneurs, however, said they intended to expand their business in the coming years.

Revealingly, although only one-quarter of the young people in Chawama were currently self-employed, the majority (77%) of youth in all employment categories indicated that they would prefer to run their own business in the future, leaving only 20% who reported preferring paid employment. These aspirations reflect the emerging positive public attitudes to entrepreneurship and the perception that employees are often badly treated and poorly paid. In fact, for a sizable proportion of the young people surveyed (38%), their current business was not their first.

Hurdles faced by young entrepreneurs

While the majority of Chawama youth believe that starting a business is their best livelihood option, young people face major challenges in establishing and running their own businesses. The most frequently cited problems are lack of capital and limited profits. Given that young people operate a limited range of income-generating activities, they face fierce competition with 60% rating it as high or very high. Consequently, they have to be inventive to attract customers: one-quarter of the young entrepreneurs seek advantages over their competitors by lowering their prices, even though this diminishes their profitability. Another quarter said they established good relations with customers, though this often implies extending favours and selling on credit, which can have a detrimental effect on the business (see also Chapter 15). Yet another quarter tried to offer higher-quality products/services or create a better product reputation.

While obtaining support from family members is one of the key ways in which entrepreneurs mobilise resources, this is not an option for all young people. For some youth, the inability or unwillingness of their family to help them set up and operate a business is frustrating. In their eyes, family support structures and norms of cooperation and trust are in decline, as indicated in the following statements from the FGDs:

> One problem we have here in Zambia is this. My brother here will not support me if I have a business idea – he will not help me with capital even if he has money. So, we don't have good support structures.
>
> (Male, 33)

> I would put it like, for example, if you want to start business, way back they just used to give [money] without even thinking of "Is he going to give it back to me or what?" People were honest. But nowadays, when you try to help a friend you will see that he runs away. You'll never see him again. There is a lack of trust.
>
> (Male, 31)

Young entrepreneurs are also limited by their lack of education, skills training, and business support services. Almost one-quarter (24%) of the youth in

Chawama had not studied beyond primary level, 30% had completed junior secondary school, 36% had completed secondary education, and just 10% had completed college or university. The general sentiment among young people was that education in Zambia was generally poor quality and of limited use for setting up and running a business. As a young man (aged 29) explained:

> When you study at school you are not told to be creating your own employment. We are just told, "When you finish school you will find employment." But they should change the syllabus like maybe trying to train a child from grade 1 to 12 to say, "I'll have a company on my own. I'll have something on my own." We don't have that but I am praying that one day, things will change so that we move in that direction.

Only 20% of the young entrepreneurs had received any skills training before setting up their businesses, although all but one thought they would benefit from undergoing additional training in how to run an enterprise. They generally considered it the government's responsibility to provide this training. Although some government programmes exist (see Chapter 4), none of the entrepreneurs had received any such support in starting or operating their business; only 3% had received some form of support from private enterprises, churches, or NGOs. This indicates that the numerous programmes initiated in recent years to spur entrepreneurship and job creation have not benefited the youth of Chawama. Almost half (46%) the young people surveyed were not even aware of the existence of any government or nongovernment entrepreneurship or enterprise development programmes and just over one-third (35%) were unaware of the National Youth Policy. Furthermore, as the following extract from one of the FGDs indicates, young people are highly critical of the availability, accessibility, and/or management of youth entrepreneurship programmes:

> In my view the government has adequate programmes. It has adequate policies for us the young people.... What is killing us is grant corruption. The levels of corruption in this country are too, too high.... They give loans up to ZMK300 million but to get that ZMK300 million there have to be kickbacks. You have to pay someone maybe half of that so what do you really have at the end of the day?
>
> (Male, 33)

Although all the youth-run enterprises could be considered to operate in the informal economy as they are not registered businesses, they are not completely shielded from governmental regulation. About one-fifth of the enterprises (21%) claimed that they were regulated by the Lusaka City Council and the majority found the regulations unaccommodating. Hence, while young people hardly receive any support from the government, some of them are still subject to its regulations, including paying fees and licences.

It has been suggested that informal entrepreneurs might benefit from organising and engaging in associations in terms of training, providing welfare services, and negotiating with the government for better conditions (Lindell, 2010). Only 15% of the young entrepreneurs, however, were members of registered business associations, such as the Market Traders' Association, or a union. Slightly more (20%) were members of informal business support groups such as *chilimba*, which are informal rotating credit and saving groups. By paying in a small amount of money on a regular basis, entrepreneurs pool their limited financial resources, which are then allocated – usually monthly – to one member of the group in turn (Chigunta, 2007).

Another factor that can greatly influence entrepreneurship in the Zambian context is the prevailing belief in witchcraft. When asked about this in the questionnaire survey, only a small proportion (9%) agreed with the statement that "using juju or charms is good for business", but belief in the power of witchcraft is widespread. Someone whose business is doing well may be accused by others of having used witchcraft to get ahead, which can act as a constraint, making people reluctant to expand. As a young man (aged 33) explained:

> The rate of progression among the businessmen worries the community. Today, for example, you have one bus, the next day we see you with 12 buses. Really, that tends to raise suspicion. So, whether you got a loan or what, people will just say "*nimfwiti, nimfwiti*", meaning he is a witch.

Reluctance to associate with successful businesspeople for fear of their potential use of charms and association with satanism reduces the use of mentors, which can otherwise be highly beneficial to young entrepreneurs.

Turning hurdles into opportunities

Despite facing the difficulties highlighted above, young entrepreneurs demonstrate a keen ability to spot opportunities where others see constraints. There are numerous examples of this happening throughout Chawama. One of the most common is dividing goods for sale into small portions to adapt to the low purchasing power of the community. This is typically the case where selling everyday items – such as rice, maize meal, salt, sugar and oil, as well as charcoal – is concerned. Other retail activities in which some young people engage operate close to the edge of the law. The brewing and sale of illegal alcohol, which both men and women engage in, is widespread in the settlement. Some young people deal in illegal substances, such as marijuana, while a minority steal goods within and outside the compound. For young women, prostitution is another avenue for generating income but due to the associated stigma it is rarely spoken about openly (Chigunta, 2007).

Some young entrepreneurs offer inventive services as and when the need arises. For example, while conducting the qualitative fieldwork, large parts of Chawama were flooded, severely hindering mobility within the compound. Standing in a strategic location close to where a section of the street was submerged in 30 cm of water, a group of young men had turned this situation of adversity – caused by the poor drainage and sewerage systems – into a business opportunity by renting out gumboots (Figure 5.1). Like many other people that day, we happily paid a small fee to borrow their boots to reach higher ground about 50 m away. Similarly, some young men fill in potholes on the roads, hoping that passing vehicles will acknowledge their work with a token payment, while others act as "traffic police" when narrow roads become jammed with cars and minivans. Although not exactly a business, one of the more ingenious activities of the youth we encountered was to become a "professional mourner". Funerals are frequent in Lusaka, given the prevalence of HIV/AIDS, and are widely attended with transport in buses and open trucks often provided for the mourners. As food and drink are always provided at funerals, some young people have taken to attending the funerals of people they do not even know as a means of obtaining something to eat for the day.

More resource-rich young people run businesses that cash in on the disadvantaged environment on a grander scale. Given the lack of nursery schools in Chawama and the large population of children and youth, some young people have turned this void into an opportunity by opening small private nursery schools. Although these are typically established in a private dwelling with large classes crowded into small rooms, they offer an invaluable service, which the state is failing to provide.

Figure 5.1 Young men renting out gumboots in flooded Chawama (photo: Thilde Langevang).

Social entrepreneurs as change makers

While some young people in Lusaka engage in dubious forms of enterprise as indicated above, others partake in activities with the specific aim of helping not only themselves but also other young people in the community. Witnessing the desperate situation of many of their peers and the evident lack of skills training, some young people have established social entrepreneurial activities to build youth capacities. One such example is the Chawama Youth Project (CYP).

Established in 2001 by a group of 10 young people, the CYP is a youth training centre run by and for young people. The decision to establish it followed the failure of many youth in Chawama to access skills training and jobs. This prompted a group of young people to try to establish a skills training centre to help disadvantaged youth in their area, though initially they failed due to a lack of resources. Their break came when the first author, then a PhD student, visited the area to conduct field research. Working through the local leadership, he recruited a group of young people to help him conduct his field research. When they complained that many researchers had visited their area and made promises to help but none had done so, Francis assured the youth that he would do his best to organise assistance for them (Figure 5.2). At the end of his fieldwork, Francis invited some of the youth to attend a workshop he had organised at the University of Zambia. There, he presented his research findings to an audience that

Figure 5.2 Francis Chigunta with peer researchers at Chawama Youth Project (photo: Thilde Langevang).

included representatives of some major local companies. Their help was sought in supporting the Chawama youth as part of their corporate social responsibility. One of the companies, BP (Zambia) Limited, responded by financing the construction of a workshop and office space, and purchasing some training equipment for the youth. The land was given to the CYP by the local council, which accepted it was unable to fulfil its role in providing training for young people and does not charge rates. This marked the beginning of the CYP.

The programme is managed by two of the original team members – young men who have persevered with the project in the face of adversity – although frustration at the continued challenges faced has forced the rest of the youth who started the project to quit. An advisory committee comprising six members, only one of whom may be considered an adult, serves as the board of directors. The CYP targets young people with limited education for training in carpentry and joinery, auto-mechanics, design, cutting and tailoring, and short computer courses. These are intended to provide the disadvantaged youth of Chawama with skills that will enable them to establish their own businesses. To identify target beneficiaries, the CYP works through churches and grass-roots organisations such as the Ward Development Community. The CYP is primarily self-funded, with its main source of income being the carpentry products made by the young people it trains, which are sold within Chawama. Young people are charged a minimal ZMK100 per month to attend the training courses regardless of the course. Five trainers are employed who are paid a low wage. Although the CYP has the capacity to train up to 320 youth annually in a variety of skills, on average limited funds have resulted in only 90–120 young people being trained. In 2013, however, the CYP received a grant of ZMK30,000 from the government's Youth Development Fund to help finance its operations, which enabled the project to operate at full capacity.

The case of the CYP illustrates young people's refusal to accept the status quo and shows how they can attempt to create social change by promoting skills acquisition that enables youth to set up a business. The project is an example of a hybrid social entrepreneurial organisation that relies on occasional donor funds and its own income-generating activities to pursue a social mission. Operating a fully self-sustaining training centre in a low-income settlement such as Chawama is extremely difficult given that the aim is to help disadvantaged young people who are unable to pay in full for the training. Depending on donor funds and grants is unreliable and funding is becoming increasingly scarce, while the demand for funding remains high. Hence, combining the two income mobilisation strategies is the best way to optimise resources and impact. The case also shows the importance of having outside connections with influential people who can leverage external resources. Vital to the CYP's operation, however, is its embeddedness in and accountability to the local community. The ability of social entrepreneurs to leverage resources from other stakeholders, while remaining embedded in their local community, has been shown to be key in other contexts (Di Domenico et al., 2010).

Conclusion

Young people growing up in Lusaka's compounds face extreme challenges in finding ways to support themselves. Since securing paid employment is difficult and often highly exploitative, it is increasingly common for young people to become entrepreneurs in a range of ways: by starting a business, seizing opportunities, and engaging in social entrepreneurship. Most of the businesses they run are based within the local community and have been established using savings or financial support from family and friends. Typically, these businesses involve small-scale trading or service provision and as many of them are similar this results in fierce competition. Despite living in a challenging environment, some young people have been shown to be highly entrepreneurial, turning adversity into advantage by setting up temporary services as and when required. As Locke and te Lintelo (2012: 792) also found in the Zambian context, young people demonstrate "astonishing perseverance in the face of adversity".

Other young people engage in activities that not only help themselves but also less fortunate youth in the compound. The CYP is a case in point, run by young people for young people, enabling youth to gain skills that will facilitate the establishment of their own businesses. Such projects are very important given the difficulty of accessing youth entrepreneurship programmes run by the government, which, the young people claim, are riddled with nepotism and corruption. With one-quarter of the youth interviewed already running a business and the remaining having expressed their desire to set up a business in the future, it is vital that government entrepreneurship programmes be made more accessible to youth living in the compounds and that social entrepreneurship projects such as the CYP are given the support they require.

Note

1 There were around ZMK4,000 to the US dollar in 2010.

References

Central Statistical Office. (2012). *Zambia – 2010 population and housing census: Population summary report*. Lusaka.

Chigunta, F. (2007). *An investigation into youth livelihoods and entrepreneurship in the urban informal sector in Zambia*. Unpublished PhD thesis, Oxford University.

Day, C., & Evans, R. (2015). Caring responsibilities, change and transitions in young people's family lives in Zambia. *Journal of Comparative Family Studies, XLVI*(1), 137–152.

Di Domenico, M., Haugh, H., & Tracey, P. (2010). Social bricolage: Theorizing social value creation in social enterprises. *Entrepreneurship Theory and Practice, 34*(4), 681–703.

Gough, K. V. (2008). "Moving around": The social and spatial mobility of youth in Lusaka. *Geografiska Annaler, 90*(3), 243–255.

Gough, K. V., Langevang, T., & Namatovu, R. (2014). Researching entrepreneurship in low-income settlements: The strengths and challenges of participatory methods. *Environment and Urbanization, 26*(1), 297–311.

Hansen, K. T. (1997). *Keeping house in Lusaka.* New York: Columbia University Press.

Hansen, K. T. (2005). Getting stuck in the compound: Some odds against social adulthood in Lusaka, Zambia. *Africa Today, 51*(4), 3–16.

Lindell, I. (2010). The changing politics of informality: Collective organizing, alliances and scales of engagement. In I. Lindell (Ed.), *Africa's informal workers: Collective agency, alliances and transnational organizing in urban Africa* (pp. 1–30). London and New York: Zed Books.

Locke, C., & te Lintelo, D. J. (2012). Young Zambians "waiting" for opportunities and "working towards" living well: Lifecourse and aspiration in youth transitions. *Journal of International Development, 24*(6), 777–794.

Maipose, G. S. (1989). Zambia's economic and unemployment problems: An analysis of underlying factors. In K. Osei-Hwedie & M. Ndulo (Eds), *Studies in youth and development* (pp. 19–39). Lusaka: Multimedia Publications.

Mair, J., & Marti, I. (2006). Social entrepreneurship research: A source of explanation, prediction, and delight. *Journal of World Business, 41*, 36–44.

Moser, C., & Holland, J. (1997). *Household responses to poverty and vulnerability: Vol. 4. Confronting crisis in Chawama, Lusaka, Zambia* (UMPP Policy Paper No. 24). Washington, DC: World Bank.

Muuka, G. N. (2003). Africa's informal sector with Zambia as a case study: A challenge to scholars to close the knowledge gap. *Journal of Business and Public Affairs, 30*(1), 50–56.

Porter, G., & Abane, A. (2008). Increasing children's participation in African transport planning: Reflections on methodological issues in a child-centred research project. *Children's Geographies, 6*(2), 151–167.

Robson, E., Porter, G., Hampshire, K., & Bourdillon, M. (2009). "Doing it right?" Working with young researchers in Malawi to investigate children, transport and mobility. *Children's Geographies, 7*(4), 467–480.

Schlyter, A. (2011). *Recycled inequalities: Youth and gender in George compound, Zambia.* Saarbrücken: Lambert Academic Publishing.

6 Youth entrepreneurship in Kampala

Managing scarce resources in a challenging environment

Thilde Langevang, Katherine V. Gough and Rebecca Namatovu

Introduction

After decades of economic turmoil and political instability, the Ugandan economy is growing rapidly with GDP annual growth rates averaging around 7% during the past decade (World Bank, 2013). New enterprises have mushroomed across the country with Kampala in particular teeming with business activity in marketplaces, along roadsides, and in neighbourhoods. Most of these businesses form part of the informal economy, which is characterised by fierce competition, limited earnings, and low productivity (Potts, 2008). Given the limited possibilities for gaining public or formal private sector employment, most young labour force entrants either set up businesses or become employed within the informal economy. As highlighted in Chapter 2, the Ugandan government is promoting entrepreneurship as the key solution to what is perceived as a mounting youth unemployment crisis.

The majority of Kampala's population live in low-income settlements characterised by poor infrastructure and inadequate services that are challenging environments in which to run a business. This chapter explores the level and characteristics of the entrepreneurial activities of young people living in one such area, Bwaise, which is located on the outskirts of Kampala. Drawing on a participatory research methodology, it examines why and how young people establish businesses in a difficult urban environment and the range of challenges they face related to resource scarcity and inadequate infrastructure. The resourcefulness of the young people, as demonstrated by their ability to switch income-generating activities as circumstances change, is shown to be key to their survival. Outlining the major hurdles young people face in sustaining and expanding their enterprises, the ways in which young entrepreneurs could potentially be supported are discussed.

Bwaise: a low-income settlement

Compared to other African countries, Uganda has a relatively low level of urbanisation with less than 15% of the population living in urban areas, though this is projected to grow rapidly in the coming years (UN-Habitat, 2012). Kampala dominates the urban scene demographically, economically, and politically. It is home to an estimated population of 1.5 million (Vermeiren, Van Rompaey, Loopmans, Serwajja, & Mukwaya, 2012) and is the base for 80% of the country's industrial and services sector firms, which generate more than half the country's GDP (Mukwaya, Bamutaze, Mugurura, & Benson, 2011). Kampala has a distinctive setting, built on and around seven hills. Low- and higher-income areas are often juxtaposed, with the latter generally located on the upper slopes and the former on low-lying land that is prone to flooding. The lower-income areas, which house the majority of the urban population, offer poor access to services, thus limiting inhabitants' daily lives and businesses (Vermeiren et al., 2012).

Bwaise, which was established in the early 1920s, is a heterogeneous residential and commercial area that is home to an estimated 40,000 people. Many of the houses have been inherited by descendants of the original owners who now live in better neighbourhoods but retain their property to rent out rooms. Consequently, many of the area's residents are tenants. Bwaise's low-lying neighbourhoods are prone to frequent flooding and have inadequate infrastructure, and most residents live in overcrowded, insanitary conditions. The higher areas offer better-quality housing and infrastructure, and the population is less dense. Bwaise has a large migrant population from all over Uganda. Many different types of businesses operate here in marketplaces, along busy roads and narrow footpaths, and from people's homes. Young people are prominent as entrepreneurs and engage in a range of activities including petty trade, food preparation, selling second-hand clothes, car washing, and hairdressing.

A broad and inclusive understanding of entrepreneurship is employed in this chapter to encompass the practices that young people engage in when seeking to make a living through starting and running businesses of various types. In analysing young people's paths to enterprise ownership in a volatile environment, we engage with the concept of "zigzagging", which has recently been employed to denote the ways in which young people in the global South often have to change lines of business and seize opportunities as they arise (Jeffrey & Dyson, 2013; Jones, 2010). To understand how young entrepreneurs deal with resource paucity and limited institutional support, we also draw on the concept of bricolage (Lévi-Strauss, 1966) – a practice that involves making do with whatever is at hand. It consists of the ability to create something out of seemingly nothing in a situation of resource scarcity and entails creatively mobilising, re-using, and combining underutilised or discarded resources that pre-exist in the environment (Baker & Nelson, 2005).

Methodology

This chapter draws on a participatory study conducted in collaboration with Uganda Youth Welfare Services (UYWS) – a youth organisation involved in a range of activities that seek to empower young people in Bwaise, including providing counselling, vocational training, and entrepreneurship education to disadvantaged youth. With the assistance of UYWS, 12 young residents of Bwaise were selected to participate in the study as peer researchers who worked with the authors over a one-month period, for which they were paid a wage and provided with lunch. The research in Bwaise combined quantitative and qualitative methods. In order to generate knowledge about the levels and types of youth entrepreneurship, a questionnaire survey was conducted with a sample of 400 young people. The wording and sequencing of the questions underwent numerous rounds of scrutiny in collaboration with the peer researchers, following which a pilot study was carried out. Prior to the data collection, the peer researchers were trained in how to randomly select respondents, administer the questionnaire, and ensure ethical conduct. In the field, each of the authors supervised a team of four peer researchers. The respondents were selected as randomly as possible by targeting every 10th household (by counting rooms) and then using random numbers to select one young person within that household to interview. Most of the questionnaires were conducted in the local language, although English was used in a few cases.

A range of qualitative methods was then employed to generate deeper insights into the practices and experiences of young business owners. Focus group discussions were conducted with four groups of young people: female entrepreneurs, male entrepreneurs, female and male entrepreneurs, and a mixture of female and male students, wage workers, and entrepreneurs. These discussions allowed us to explore young people's understanding of the concept of entrepreneurship as well as the challenges and opportunities facing young entrepreneurs, the resources they draw on, and the institutional support and/or barriers to entrepreneurship they face. To stimulate reflection on the meaning of "entrepreneurship" and "successful entrepreneurship", photographs displaying local entrepreneurs of various kinds were used. Drawing inspiration from Participatory Rural Appraisal (PRA) techniques, a ranking and scoring system was introduced to determine the key factors that, in the participants' views, inhibit or promote entrepreneurship.

In-depth interviews were conducted with 20 business owners, using an open-ended approach that allowed them to raise issues they considered important. They were asked to narrate their life stories, highlighting their work histories and key turning points in their lives, why they decided to start a business, how their businesses had changed over time, the challenges they faced, and their aspirations. Furthermore, two short videos were made by young people in Bwaise to enable them to tell their own stories of youth entrepreneurship (Figure 6.1). The Kampala branch of Slum Cinema was brought in to teach the young people how to use a video camera, write a script, use images to tell a story, film their own story, and edit and screen it. Working in two groups, 20 young people were involved in making these films; one made a film on business innovation and imitation, and the other on why many micro-enterprises fail to grow and often have to change their line of business (see http://vimeo.com/72811329).

Figure 6.1 Video-filming by youth in Bwaise (photo: Thilde Langevang).

During the data-collection process, we found that many business owners had not received any business training but were interested in doing so. Subsequently, we organised a tailor-made training session in partnership with the Entrepreneurship Centre at Makerere University Business School, which is experienced in training micro-entrepreneurs in their local language. Since the training was interactive, it also contributed to our research findings as new issues arose during the discussions. At the end of the research process, a dissemination workshop was held during which the study's preliminary findings were discussed with key stakeholders and some of the research participants (for more details, see Gough, Langevang, & Namatovu, 2014).

Convoluted paths to enterprise ownership

The questionnaire survey conducted in Bwaise revealed that the most frequent employment status among young people was entrepreneur (27%), followed by informal wage employment (18%) and formal wage employment (11%). Twenty-eight per cent of the young people were students who did not work on the side and 16% said that they were unemployed or homemakers (see Table 6.1). More young men (34%) than young women (20%) were entrepreneurs.

Most of the young people in our sample of entrepreneurs (72%) had been employed in the informal sector prior to starting their business; only a few (7%)

Table 6.1 Employment status of young people in Bwaise

Employment status	Percentage
Entrepreneur	27
Informal wage employment	18
Formal wage employment	11
Students (not working)	28
Unemployed or homemakers	16
Sample size	*400*

Source: YEMP questionnaire (2011).

had started their business following formal wage work. The remaining 21% had started a business after leaving school, being unemployed or fully engaged in household work. This shows how most young people gain work experience before setting up their own business. The vast majority of businesses run by young people are in the services and retail sector (90%) with only a few in the manufacturing sector (6%) or farming (4%). The most common businesses are petty trade, small shops selling foodstuffs, food preparation and sale, hairdressing, carpentry, *boda boda* (motorbike taxis), and trading new or second-hand

Figure 6.2 Peer researcher interviewing shop owner in Bwaise (photo: Thilde Langevang).

shoes and textiles. The majority (61%) of these businesses were located in Bwaise, one-third elsewhere in Kampala, and 6% outside Kampala.

Respondents' stories revealed complex and convoluted paths to business ownership: John's tale illustrates how many young people face numerous hurdles and changes of direction along the way. John started working at the age of 15 when, as an orphan, no one in his family could pay for him to continue his studies. He started dealing in fuel – an illegal activity that involved siphoning off diesel from vehicles which he then exchanged for paraffin and sold in the police barracks where he lived with his brother in Lira, northern Uganda. At the age of 17, John moved to Kampala, initially living with his sister in the neighbourhood of Namasuba. He learnt barbering from a friend, needing just a week to pick up the necessary skills, following which he worked for other people in their salons, giving the owner a share of his daily earnings. In an attempt to increase his income, he would listen to the radio to identify the more expensive salons and target them. John moved to Bwaise at the age of 20 when a room in his grand-mother's house became available. Becoming frustrated with the little money he was earning from barbering, he joined an Indian construction company as a painter, learning the trade on the job. Despite liking the work, he left after two years as he could not afford to gain a qualification that would have enabled him to be better paid, rather than casually employed on a daily basis. John then set up on his own as a builder, offering to carry out general construction work as well as painting and electrical wiring. Although he claimed that business was good when he had a job to do, he was without work half the time, which he attributed to people preferring to employ someone older and more experienced.

On visiting a sister living in Entebbe, located on Lake Victoria, he decided to take the boat to Tanzania to try his luck there. En route, he made a friend who ran a garage in Masaka and was invited to work for him as a mechanic since he had some knowledge of engines. After five years in Tanzania, John found he missed home so he returned to Kampala, setting up in the building trade once again. To supplement this work, he started selling sand extracted from the storm drain in front of his home. Subsequently, this became his main business. When asked if it was a good business, John, now aged 27, replied that it was not as he did not get customers every day. He indicated, however, that a side benefit of his work was that it helped keep the storm drains clear because it involved extracting both the sand as well as the garbage mixed in with it, which reduced the chance of flooding. John's future plans were to set up a shop selling building equipment on the out-skirts of Kampala but he was being hindered by lack of capital.

The spatial metaphor of "zigzagging" (Jeffrey & Dyson, 2013; Jones, 2010) is fitting in this case. As John's story shows, young people meander through an uncertain environment, constantly adapting their livelihood activities to shifting circumstances and chances that present themselves in the moment. They seize opportunities as they arise and change their path when new constraints or risks emerge (Langevang, 2008). Zigzagging may thus include both illegal activities in the shadow economy as well as a wide range of legitimate economic activities located in what is commonly termed the informal economy.

The majority of young people who had started a business following informal wage work felt that it was far better to own a business than to be informally employed by somebody else. Many had felt exploited by their former employers and complained about earning meagre sums. As Paul, the 19-year-old owner of a small shop, said when asked why he had decided to start his own business: "I did not want to be used. If you are employed by somebody else, they pay you less and they mistreat you." For many young people facing limited formal employment opportunities and finding informal employment exploitative, the option of starting an enterprise was preferable. In the words of Jenifer, a 30-year-old laundress who had previously worked in a restaurant,

> I got fed up because the work was so hard, but the pay so little. They would make us work so much but the pay was tiny ... I [had] failed to get [any] other kind of employment and decided to start this business.

Some young people also emphasised the learning and independence gained through self-employment as a key motivating factor in setting up a business. Lameck, aged 24, who owns a video hall together with his brother, observed that self-employment

> teaches responsibility. Nobody is following up on you. You have to be there and be good with the customers. Secondly, the freedom. If you are employed, you have to come in everyday even if you are sick or if you have to go for a wedding, funeral or something in the village. When you are self-employed, you can decide when to come and go. You are the one in control.

Young people's stories also revealed that their reasons for pursuing a business are closely connected with a desire not only to improve their own situation, but also the well-being of their families. Echoing Khavul, Bruton, and Wood (2009), who find that expectations of – and obligations towards – the extended family are paramount in East Africa, many young people feel obliged to shoulder the responsibility of looking after their families; starting a business is seen as the best route to doing this. Furthermore, several young men and women claimed that they hoped to be able to create a better future for their children through the businesses they were now running. Being embedded in an environment characterised by high levels of poverty, many young people had a strong desire to help others in the community through their businesses. Paddy, for example, who produces handicrafts and artwork, does not employ anyone but trains other young people free of charge and encourages them to become self-employed. "I advise young people to join the handicrafts sector since it is cheap to start. They should know that the government has no jobs", he said. Janet, aged 26, who has managed to set herself up as a seamstress – despite being challenged by a speaking and walking disability – teaches unemployed girls how to sew. As she explained, "Most of the girls around here don't have work, so I think it is better if they get a skill."

Managing scarce resources in enterprise start-up and operation

Most young business owners started their business with small amounts of capital derived from either their own savings (47%) or their family (46%). A few had received support from friends (7%) or informal moneylenders (1%). Notably, none had taken out loans from formal financial institutions such as microfinance organisations or banks. This shows how, in an environment characterised by extreme resource scarcity, young entrepreneurs' business start-ups are closely intertwined with the resources available to them through their social networks. Many young people had started a particular line of business on the advice of friends or family and depended heavily on these networks to acquire not only capital, but also skills, motivational support, information, business advice, and access to suppliers and customers.

In terms of vocational and entrepreneurial skills training, only around one-third had received any form of formal training before starting their business, whereas one-fifth had received training after having established their business. The vast majority (91%) said they had not received any form of institutional business support. The 9% who had obtained support had done so from private businesses, nongovernment organisations, or religious organisations. Revealingly, none had received any support from the government and more than half (55%) were unaware of the existence of government or nongovernment support programmes for youth or enterprise owners, despite there being numerous examples of both (see Chapter 2). Moreover, only 6% were engaged in formal business associations. Consequently, young entrepreneurs have to manage with the limited resources available to them in their immediate surroundings and networks, and engage in practices of bricolage. With very limited financial resources at hand, and often lacking vocational and business skills, young people have to take the little they have and know as their point of departure for setting up a business. Jenifer, for example, started a small laundry business washing clothes by hand because, as she said, "It is the only thing I could do." With her experience of washing clothes from home, the business required very limited investment – the purchase of a large washing bowl.

Other young people were involved in the creative reuse of waste material, such as scrap, garbage, paper, and second-hand clothes and fabric that can be acquired free or at low cost. A young man who had participated in one of the videos made curtains out of second-hand sheets; by decorating them with colourful paint and adding curtain hooks, he was able to add considerable value to otherwise discarded products. Paddy told how he had started his handicraft business from "nothing", making use of "old pieces and materials" to construct novel artwork. He drew motivation from his passion for art, making use of some handicraft skills he had acquired from a short course organised by a religious organisation. Paddy's business had expanded considerably from its humble beginnings and he now produced a variety of items, including colourful necklaces and bracelets made from waste paper. Another way to make the most of the resources at hand is to use the body as capital (Esson, 2013). This is the case for young people who use their bodily strength to carry goods or water, and also for the young women in Bwaise who engage in commercial sex. Other young people, especially young men, have become experts in more dubious forms of business, such as dealing in drugs, theft, and selling stolen goods.

Being an entrepreneur in a challenging environment requires flexibility and the ability to respond to changing opportunities and constraints. Almost a third (30%) of the young people we met had already changed their line of business once or several times. Common reasons for doing so included: lack of profit because of high competition, high expenses and/or high taxes, perceiving a more promising opportunity, and changes in family circumstances such as marriage or giving birth. The latter mostly affects women, who are often expected to stay at home doing the housework and taking care of the children following marriage and childbirth. While this does not necessarily entail an end to their entrepreneurial career, women have to find a business that is compatible with their role as homemakers, often resulting in them setting up businesses they can operate from home (Gough, 2010).

Another common practice employed by youth to deal with a volatile environment and make the most of scarce resources is to operate several businesses at the same time: 15% of those interviewed in Bwaise ran more than one business. The case of Helen, aged 25, is illustrative. Helen started by setting up a hairdressing salon in a shack but has since added several businesses: she produces and sells passion-fruit juice, manages a mobile phone service stall, and operates a pawn shop through which she lends small amounts of money to neighbours at a low interest rate. Young people diversify for a range of reasons: to spread risks so that if one business fails there is the chance that another will succeed; to ensure a steady income throughout the year in seasons where one business is bringing in little or no profit; to generate additional income to cover a new responsibility, such as paying school fees for a sibling; and for the prestige associated with having several businesses as this indicates success and the ability to manage resources (Langevang, Namatovu, & Dawa, 2012).

Hurdles faced by young entrepreneurs

Although young people are resourceful in carving out livelihoods with limited means, they face immense challenges in sustaining and expanding their businesses. Participants of the focus group discussions agreed that there were few or no successful young entrepreneurs in Bwaise. They associate successful entrepreneurship primarily with the ability to expand a business despite starting small and having few resources. As one focus group participant explained, "Generally, young people are not successful, they are just on their way", and similarly another noted, "The young ones are just trying – they are not yet in the bracket of successful." The findings from the questionnaire survey support these statements, revealing that when growth is measured in terms of the number of employees, young people's businesses tend to remain at the micro-level (see

Table 6.2 Number of employees in youth-run enterprises in Bwaise

Number of employees	Percentage
0	45
1	32
2–5	16
>6	7
Sample size	400

Source: YEMP questionnaire (2011).

Table 6.2). Almost half of the entrepreneurs operate their business on their own, about one-third have just one employee, 16% have two to five employees, and 7% have more than six employees. The majority of employees are family members (42%) or casual workers (39%), with only 11% employing permanent workers. The remaining 8% are apprentices.

Respondents claimed that a range of factors inhibit their growth. A key challenge is the lack of skills; two-thirds of the entrepreneurs stated that they lacked business education and wished to undertake training in entrepreneurship and business management, and in all four focus group discussions, inadequate education and training ranked among the top three challenges. Another problem that was often raised concerned the regulatory environment: 60% of the business owners believed that the regulations to which they were subject by the local council hindered their businesses to a "large" or "very large" extent. In particular, the payment of licences and taxes, and the associated lack of transparency, were pinpointed as problems. While some entrepreneurs said they managed to hide from the tax collectors, many complained that they were subject to corrupt practices. They also found it unfair that new businesses should have to make payments before it became clear whether they were profitable. The existence of regulation in terms of granting licences and collecting taxes indicates that, while these enterprises would popularly be termed "informal", many are in fact subject to regulation by the local government.

Another major hurdle facing the vast majority of entrepreneurs in Bwaise relates to the area's poor infrastructure: the most common obstacles to businesses include garbage (98%), flooding (93%), lack of sanitation (90%), power shortages (87%), and water shortages (72%). While many young people are quite content living and running businesses in Bwaise – emphasising the large customer base and relatively low cost of living – the inadequate infrastructure remains a real burden. As Mary (aged 28) stated, "This place is just floods and garbage, no good thing – look at our living conditions!" She pointed to the flooded area around her – a brown mix of rainwater and sewage adjacent to a pile of garbage where marabou storks were feeding. For business owners, the frequent flooding poses a high cost in terms of ruining business premises and stock, blocking customer access, and taking up much time and energy to clear (see Figure 6.3).

Figure 6.3 Scrap collector in front of flooded business premises in Bwaise (photo: Thilde Langevang).

Competition is another key concern for young entrepreneurs who tend to set up similar businesses. Although some try to make their products and services unique, there is clearly a limit to their inventiveness. Dealing with customers was another challenge cited in the focus group discussions and in-depth interviews. Since the majority of businesses serve mainly local residents, who for the most part have very low and irregular incomes, buying on credit is a common practice and many entrepreneurs complained about the prevalence of bad debtors and deceitful customers. Moreover, many young entrepreneurs believe it is more difficult for them to acquire customers than for older people, given the tendency to perceive youth as being less experienced and less trustworthy. Accessing capital for expansion is yet another key hurdle facing young entrepreneurs. The vast majority of young people rely on their own savings or social networks, in particular, family and friends. While 13% had attempted to borrow from a microfinance institution none had succeeded, despite claims by such institutions that they cater to the poorest and most vulnerable (see Chapter 16).

Although social networks are key assets for young people, they tend to be concentrated in a narrow circle of friends and family whose own resources are likely to be constrained. This means there is a limit to the support young entrepreneurs can obtain. As studies from elsewhere show, relying mainly on this form of bonding social capital through strong ties may enable entrepreneurs to "get by" but rarely to "get ahead" (Turner & Nguyen, 2005; see also Chapter 15). Moreover, social networks can also concomitantly act as constraints: if one young person's business starts to thrive, others within the network expect to receive support, hindering the former's ability to invest in his or her own business. As others note, the obligation to cater for less fortunate family members often falls on entrepreneurs (Khavul et al., 2009), with their businesses sometimes becoming rather like "relief organisations" (Kuada, 2009). While sometimes lamenting that they were unable to expand their businesses because of such obligations, young people did not question their responsibilities. Paul, who had migrated to Kampala aged 15 to start working to support his family back in the village, stated,

> Sometimes I do feel that it is a big responsibility, you may even say a burden, but I am the only one to support them and it makes me proud to see that they are not suffering as much as before.

As Imas, Wilson, and Weston (2012: 577) also find in the case of entrepreneurs in Accra, family is a key "means of being". Accordingly, when entrepreneurs are not able to cater for their families because of a failed business or fluctuating income, they become embarrassed and frustrated. Several young entrepreneurs had been forced to leave their spouse and children in the village or to send their children to live with other family members simply because they were unable to look after them financially.

Another contentious issue worrying young people is the practice of witchcraft. Many entrepreneurs believe their competitors use witchcraft in order to get ahead and that the more they engage in witchcraft the more successful their business is likely to be. Given its contentious nature, only 16% of the questionnaire respondents said that witchcraft was good for business, but during the training sessions participants voiced the view that witchcraft did indeed influence business and was needed to beat one's competitors.

Exposed to so many challenges – often simultaneously – makes operating, let alone expanding, a business extremely difficult for young people. Despite the many hurdles encountered, the vast majority of young entrepreneurs (91%) aspired to expand their business. Their ideas included expanding their business premises or acquiring better-located premises, obtaining more advanced technology, adding a business, or changing their line of business. Few entrepreneurs, however, had clear plans for how they were going to achieve these aspirations and expected many hurdles along the way. Simon, a 25-year-old retail shop owner, said he wanted to become a wholesaler, but explained that he was reluctant to make any plans because he had already been disappointed numerous times in trying to expand his business. Despite this, he was determined to carry on with his efforts.

Conclusion

This chapter has illustrated the entrepreneurial practices of young people operating small businesses in a very difficult urban environment. While they have started their businesses for a range of reasons, improving their own welfare and that of their families is a key motivating factor. Operating in a business environment characterised by a plethora of challenges and limited institutional support, and possessing very scarce resources, entrepreneurs in Bwaise manage by drawing on their ingenuity and creativity. The concept of bricolage captures well the ways in which young people make do with limited resources – re-using waste or scrap material, relying on the body as their key asset, diversifying business activities and/or changing lines of business, and drawing on the available resources in their networks. Having to constantly adjust their livelihood activities to the changing opportunities, responsibilities, constraints, and risks in the environment means that young people's path to enterprise ownership and their progression into business is by no means linear; they "zigzag" their way forward.

While documenting young people's entrepreneurial skills and resourcefulness, this chapter has also highlighted how very few young entrepreneurs are "successful" when measured by their own yardsticks of growth. Their lack of progress can be linked to a range of interconnected challenges related to poverty and limited institutional support, including lack of access to education and capital, poor infrastructure, and no transparency in local authority regulation. This shows how the government has largely left Bwaise's youth to fend for themselves despite its rhetoric of development and its aim to ease the mounting youth unemployment crisis by supporting young entrepreneurs. Both the young people, as well as the stakeholders that participated in the dissemination workshop, voiced their discontent over the lack of support mechanisms for youth in Bwaise and had numerous suggestions as to how the situation for young entrepreneurs could be improved. The need for business education and training was one area that emerged as being key to this. Providing young people with basic knowledge of business management is, therefore, a potential support mechanism that needs to be explored. Since conventional formal classroom-based teaching might not work for many young people, there is a need for innovative content and delivery methods that suit their needs better.

Improving young people's knowledge of financial institutions and developing youth-friendly loans and savings schemes is another important area where support to young entrepreneurs could be improved. Establishing mentorship programmes and developing business networks, through which young people are linked to more experienced entrepreneurs who can provide advice and encouragement, were also identified as useful support mechanisms. Key to improving the business environment for young entrepreneurs in Bwaise, however, is improving its infrastructure. This highlights how a holistic approach to business support is necessary – one that goes beyond classic entrepreneurship promotion tools and also addresses the services fundamental to running a business.

References

Baker, T., & Nelson, R. E. (2005). Creating something from nothing: Resource construction through entrepreneurial bricolage. *Administrative Science Quarterly, 50*, 329–366.

Esson, J. (2013). A body and a dream at a vital conjuncture: Ghanaian youth, uncertainty and the allure of football. *Geoforum, 47*, 84–92.

Gough, K. V. (2010). Continuity and adaptability of home-based enterprises: A longitudinal study from Accra, Ghana. *International Development Planning Review, 32*(1), 45–70.

Gough, K. V., Langevang, T., & Namatovu, R. (2014). Researching entrepreneurship in low-income settlements: The strengths and challenges of participatory methods. *Environment and Urbanization, 26*(1), 297–311.

Imas, J. M., Wilson, N., & Weston, A. (2012). Barefoot entrepreneurs. *Organization, 19*(5), 563–585.

Jeffrey, C., & Dyson, J. (2013). Zigzag capitalism: Youth entrepreneurship in the contemporary global South. *Geoforum, 49*, R1–R3.

Jones, J. L. (2010). "Nothing is straight in Zimbabwe": The rise of the Kukiya-kiya economy 2000–2008. *Journal of Southern African Studies, 36*(2), 285–299.

Khavul, S., Bruton, G. D., & Wood E. (2009). Informal family business in Africa. *Entrepreneurship Theory and Practice, 33*(6), 1217–1236.

Kuada, J. (2009). Gender, social networks, and entrepreneurship in Ghana. *Journal of African Business, 10*(1), 85–103.

Langevang, T. (2008). "We are managing!" Uncertain paths to respectable adulthoods in Accra, Ghana. *Geoforum, 39*(6), 2039–2047.

Langevang, T., Namatovu, R., & Dawa, S. (2012). Beyond necessity and opportunity entrepreneurship: Motivations and aspirations of young entrepreneurs in Uganda. *International Development Planning Review, 34*(4), 439–460.

Laws, S., Harper, C., & Marcus, R. (2003). *Research for development: A practical guide.* London: Sage.

Lévi-Strauss, C. (1966). *The savage mind.* Chicago, IL: University of Chicago Press.

Mukwaya, P., Bamutaze, Y., Mugarura, S., & Benson, T. (2011, May). *Rural–urban transformation in Uganda.* Paper presented at the IFPRI–University of Ghana Conference on "Understanding Economic Transformation in Sub-Saharan Africa", Accra.

Porter, G., & Abane, A. (2008). Increasing children's participation in African transport planning: Reflections on methodological issues in a child-centred research project. *Children's Geographies, 6*(2), 151–167.

Potts, D. (2008). The urban informal sector in sub-Saharan Africa: From bad to good (and back again?). *Development Southern Africa, 25*(2), 151–162.

Robson, E., Porter, G., Hampshire, K., & Bourdillon, M. (2009). "Doing it right?" Working with young researchers in Malawi to investigate children, transport and mobility. *Children's Geographies, 7*(4), 467–480.

Turner, S., & Nguyen, P. A. (2005). Young entrepreneurs, social capital and Doi Moi in Hanoi, Vietnam. *Urban Studies, 42*(10), 1693–1710.

UN-Habitat. (2012). *Uganda national urban profile.* Nairobi.

Vermeiren, K., Van Rompaey, A., Loopmans, M., Serwajja, E., & Mukwaya, P. (2012). Urban growth of Kampala, Uganda: Pattern analysis and scenario development. *Landscape and Urban Planning, 106*(2), 199–206.

World Bank. (2013). *Uganda economic update: Bridges across borders: Unleashing Uganda's regional trade potential.* Washington, DC: World Bank.

7 Prospects and challenges of youth entrepreneurship in Nima-Maamobi, a low-income neighbourhood of Accra

Paul W. K. Yankson and George Owusu

Introduction

Over the last three decades public sector downsizing as a result of economic reforms and liberalisation has led the manufacturing sector to shrink with the collapse of uncompetitive firms. This has reduced the opportunities available for wage employment in the formal economy (Baah-Boateng & Turkson, 2005), which has affected Ghana's cities and other large urban centres where the bulk of formal sector wage employment opportunities are usually found. It is also against the backdrop of rapid urbanisation which Ghana has experienced since the 1950s. In 1921, when the country's population was only 2.3 million people, only 8% lived in urban areas. This proportion increased to 23% in 1960 when the national population was 6.7 million. Post-1960, the population doubled to about 12 million inhabitants in 1984 and again to 24.6 million in 2010; in the same years, urbanisation levels increased from 31 to 51% (Ghana Statistical Service, 2012).

This trend in urbanisation has affected unemployment and underemployment in urban areas. Although poor data make it difficult to determine the extent of urban unemployment in Ghana, the Institute of Statistical, Social and Economic Research (ISSER) (2010, 2013) indicates that the youth cohort is worst affected.[1] Youth unemployment in Ghana is a major concern in both urban and rural areas but is more pervasive in urban areas, given the pressure on the urban labour market as a result of the high rate of rural-to-urban migration among young people combined with the natural increase in urban populations. Informal low-income settlements in particular are characterised by high levels of unemployment among youth due to their limited education and weak socioeconomic backgrounds. These same settlements are also sites for vibrant entrepreneurial activity, however, especially street- and home-based economic activities (Gough, 2010). As is the case elsewhere in sub-Saharan Africa, most young people end up operating within the informal economy, either working for others or setting up and running their own enterprises (Sommers, 2010).

The youth employment challenge has featured prominently in various government and nongovernment policies and programmes, although the overall approach has been to create employment opportunities for youth, including self-employment. The thrust of existing government policy, as highlighted in the current medium-term policy framework – the Ghana Shared Growth and Development Agenda – is youth entrepreneurship as a response to youth unemployment (Ghana National Development Planning Commission, 2010). Although entrepreneurship is not youth-specific, the widely held view is that it can unleash young people's economic potential and provide them with livelihood opportunities (Kararach, Hanson, & Léautier, 2011). Drawing on Nima-Maamobi, a low-income settlement in Accra, as a case study, this chapter assesses the opportunities and challenges confronting youth entrepreneurs in an urban environment. Specifically, the chapter highlights the background characteristics of youth entrepreneurs and the enterprises they run, examines youth enterprise formation – including their skills acquisition, social networks, and other inputs required to run a business – and assesses the opportunities and challenges confronting young people and how these affect their business operations.

Nima-Maamobi: a low-income settlement

Compared to many sub-Saharan African countries, Ghana is highly urbanised with just over half of the total population living in urban centres in 2010 (Ghana Statistical Service, 2012). The urban population is concentrated in a few centres, however, with Accra and its immediate surrounding districts – usually referred to as the Greater Accra Metropolitan Area – accounting for over 30% of the total urban population. The city of Accra has been the capital of Ghana since 1877 when the British colonial administration transferred the capital from Cape Coast, attracting both public and private investment as well as an influx of people drawn to the numerous socioeconomic and other livelihood opportunities the city offered (UN-Habitat, 2009). Accra's population increased from just under 400,000 inhabitants in 1960 to almost 1.9 million inhabitants in 2010. Though migration is still an important component of population change in Accra, natural increase is the dominant growth factor with a little more than half (54%) of the city's residents born in Accra according to the 2010 Population and Housing Census.

Accra's population, like that of other African urban centres, is predominantly young; about 31% of the city's population in 2010 was aged under 25 and 42.4% was aged between 15 and 34, thus falling into the youth category. The gross population density of the Accra Metropolitan Area (AMA) is 10.03 persons per hectare as compared to 6.23 per hectare in 1970 (UN-Habitat, 2009). This has placed considerable pressure on the city's heavily backlogged housing stock, resulting in the proliferation of so-called "slums", where 38.4% of the city's population live (UN-Habitat, 2011). Finding a job in the city is also highly problematic, and over half (51.2%) of the economically active population aged 15 or above in the AMA is self-employed, either with or without employees (6.8% and 44.4%, respectively) (Ghana Statistical Service, 2012). In terms of occupation, sales and service workers accounted for 36%, followed by craft and related trade workers (18.7%) (Ghana Statistical Service, 2012).

Nima-Maamobi is situated on the northern edge of the central business district of Accra. It is a well-known low-income residential area, which emerged in the 1940s as a pasturing field for cattle prior to their sale in Accra markets (Owusu, Agyei-Mensah, & Lund, 2008). The establishment of an American military base close to Nima during the Second World War caused an upsurge in the influx of people, mainly due to the job opportunities it offered. Nima-Maamobi's population was further boosted by Gold Coast troops returning from the Second World War. Subsequent migrants came primarily from northern Ghana and other West African countries. Nima-Maamobi has developed in an unplanned manner. Although the area was declared in need of redevelopment in the 1950s, it was not until the 1990s that serious efforts were made to improve road access and other basic services. It is a densely packed, poor residential area with little room for expansion, and characterised by haphazard development, poor sanitation and drainage, and poor accessibility (Figure 7.1). Despite its low status and heavy congestion, the area continues to attract migrants from other parts of Ghana and beyond, especially the Sahelian West African region. Unlike some parts of the AMA, where the population has declined in recent years due to congestion and gentrification, Nima-Maamobi has continued to experience very rapid population growth – from 118,856 in

Figure 7.1 Narrow pathway between houses in Nima-Maamobi (photo: George Owusu).

2000 to 144,120 in 2010 – exacerbating the congestion and pressure on the settlement's limited infrastructure and services.

The population is highly heterogeneous with the following ethnic groups: Akans (24.6%), Ewe (16.3%), Mole-Dagbon (16.1%), Ga-Dangme (11.8%), and other minority groups, including Sahelian West Africans (31.2%) (Owusu et al., 2008). Due to the presence of many Muslims, there are numerous mosques in the locality. Nima-Maamobi is characterised by vibrant economic activity with many different types of businesses operating from homes, in alleys, around and in the traditional market area, as well as in shops and stalls along the main road linking it with the surrounding neighbourhoods and the rest of Accra.

Methodology

The fieldwork for this study was carried out in 2012 and included a reconnaissance visit to Nima-Maamobi to become familiar with the residential area and broadly ascertain the community's physical, social, and economic structure. The authors also interacted with officials of the Youth Development Unit at the East Ayawaso Sub-Metropolitan Assembly Office of the AMA (under whose jurisdiction Nima-Maamobi falls). Following this visit, a draft structured questionnaire targeting the youth of Nima-Maamobi and their economic activities was prepared, based on questionnaires used in similar YEMP studies of low-income settlements in Kampala (Bwaise), Uganda, and in Lusaka (Chawama), Zambia. Guidelines for conducting in-depth interviews of key informants and focus group discussions (FGDs) were also prepared. The fieldwork was undertaken in collaboration with the Youth Development Unit, whose youth coordinator helped recruit and train young people in Nima-Maamobi to conduct the questionnaire survey. The questionnaire was pre-tested and then finalised based on the feedback received. The involvement of youth (both male and female) in the field survey was a learning experience for those involved and greatly facilitated access to the informants in a settlement that can otherwise be difficult for outsiders to enter.

For survey purposes, Nima-Maamobi was divided into four sectors with field interviewers assigned to each. The respondents were randomly selected though the field enumerators were instructed to ensure that the interviews were spread across each sector spatially and that both male and female youth were sampled. In all, a total of 480 questionnaires were administered to youth in the community. The questionnaire covered such themes as: who the young people were and what was their role in the community; the activities in which they were involved and the formation of youth businesses; their skill base and acquisition process; business opportunities, expectations, and dynamics; the attitudes of society and youth towards entrepreneurship; and their knowledge of and access to support programmes for youth entrepreneurship run by the state and civil society generally – such as non-profit organisations, community-based organisations (CBOs), and religious organisations. Two FGDs, one involving young men and the other young women, were conducted at the East Ayawaso Sub-Metropolitan Assembly

Office in Nima-Maamobi. Each group comprised 10 young people aged 22–26. In addition, key informant interviews were conducted with the youth coordinator of the East Ayawaso Sub-Metropolitan Assembly, an official of the Federation of Youth Clubs, a local nongovernmental organisation (NGO), and an official of the National Council of Muslim Chiefs (who is also in charge of an NGO that focuses on imparting employable skills to youth).

Opportunities for youth entrepreneurship in Nima-Maamobi

Before turning to consider the businesses established by young people in Nima-Maamobi, this section presents a profile of the youth living in the settlement. Nima-Maamobi is still primarily a migrant community with 55% of young people having moved there compared with 45% who were born there. Given their young age, many youth migrated as minors accompanying or joining their parents or other relatives who had already settled in the city. It is only after coming of age or having secured a permanent residence that the search for employment becomes a key priority. Hence, the largest proportion of young migrants (43%) moved to Nima-Maamobi for family reasons, followed by employment/job-seeking (38%), and education (15%). This differs from most migration studies on Ghana where work/employment is the primary motive for migrating (Awumbila, Owusu, & Teye, 2014).

The highest level of educational attainment among the youth of Nima-Maamobi varies considerably, although it is generally low. As Table 7.1 shows, almost one in 10 youth had no formal education whatsoever and a further 52% had attained only basic education, i.e. primary and junior high school or middle school. Just over one-quarter (27%) had completed secondary school. Recently, the quality of Ghana's education system has been questioned, given the continuous fall in the country's standard of education, even at the secondary and tertiary levels.[2] For youth from poor communities, having no or only a low level of education is likely to compromise their life progression chances and particularly their prospects of employment. Young people who have not completed school

Table 7.1 Highest level of education attained by the youth of Nima-Maamobi

Education level	Percentage
None	10
Primary	14
Junior high/middle school	38
Vocational/commercial	5
Secondary	27
Training college	2
Tertiary (university/polytechnic)	5
Sample size	480

Source: YEMP survey (2012).

may enter the employment sphere by acquiring skills through the traditional apprenticeship system. This is especially so in the case of artisanal trades where the individual, after training, becomes a self-employed master artisan, who might then engage others as apprentices or work with other master craftsmen in a shared arrangement. Ghana's changing socioeconomic and political situation, however, has meant that the traditional apprenticeship system is also undergoing a transformation in terms of its processes and institutions and, consequently, the system now faces new challenges (Anokye & Afrane, 2014).

The current employment status of young people in Nima-Maamobi tends to vary (Table 7.2). Notably, almost half (46%) the young people claim to be unemployed, but this does not mean that they are doing nothing. Many such youth are in fact engaged in homemaking, irregular piecework, or illegal activities including prostitution, robbery, and Internet fraud (known locally as *sakawa*). Youth employment in the formal sector (the public and private sectors, international organisations, and NGOs/CBOs) is relatively limited, accounting for only 13% of the young people surveyed. A further 11% are informally employed. A significant proportion of young people (29%) run their own business and it is they who are the focus of this chapter. More than half (53%) of these young entrepreneurs were female compared with 47% who were male. While on a global scale young men tend to be more entrepreneurial than young women, the predominance of young female entrepreneurs in Nima is in line with the Global Entrepreneurship Monitor (GEM) findings for Ghana as a whole (see Chapter 3 and Langevang, Gough, Yankson, Owusu, & Osei, 2015, for a discussion of why women dominate entrepreneurship in Ghana).

Given that it is difficult to obtain formal employment, especially with education levels being as low as they are in poor urban communities like Nima-Maamobi, entrepreneurship ventures in the form of self-employment become a critical tool in the pursuit of social mobility among youth. According to ISSER (2012), youth entrepreneurship provides an opportunity for young people to be creative, to learn to manage risks, and, importantly, to develop businesses that can employ themselves and others in a world of increasingly high unemployment.

Table 7.2 Current employment status of young people in Nima-Maamobi

Current employment status	Percentage
Public formal sector	6
Private formal sector	6
NGO/CBO	1
Self-employed	29
Informal sector	11
Other	2
Unemployed	46
Sample size	*480*

Source: YEMP survey (2012).

Table 7.3 provides the main reasons young people gave for starting a business in Nima-Maamobi, by gender. About 29% of respondents cited lack of employment, but this was higher in the case of women (almost 33%). Beyond this, young people's desire to be independent (25%) and to accumulate wealth (22%) explains why they have chosen to start a business in Nima-Maamobi.

Although young people in Nima-Maamobi engage in a range of businesses, by far the most common is trading – the key activity of about 61% of the entrepreneurial youth surveyed. This is followed by dressmaking or tailoring (26%), carpentry (7%), auto-mechanic services (5%), and others (2%). Youth involvement in trading comes in different forms, such as street hawking food and non-food items (especially mobile phones, accessories, and telecom reload cards), trading from a tabletop stall, and operating from a fixed stall or shop.

The popularity of trading is due to a number of factors including: the low level of start-up capital required with the possibility of obtaining goods on credit and repaying this after sale; the limited skills and training required; and lower overhead costs and the general flexibility of business operations, among others. In addition, trading as an economic activity in Nima-Maamobi is greatly facilitated by Nima market, sometimes referred to as the ECOWAS market because it attracts cross-border traders from the West African sub-region, some of whom are already permanently settled in the community.[3] The market is, however, heavily congested, and trading activities spill onto the streets.

While dressmaking and tailoring is a common activity for young people in Ghana (see Langevang & Gough, 2012), the Nima-Maamobi area is especially known for traditional Ghanaian embroidery for both men and women. As a key informant noted, the Muslim community of Nima-Maamobi has a number of social and religious occasions (such as weddings and naming ceremonies), many of which require devoted worshippers of the Islamic faith to dress in traditional attire. This provides a good market for traditional Ghanaian apparel. However, as Langevang and Gough (2012) note, the sector faces challenges from the import of used and new clothes, and the inability of the local textile and clothing industry to respond adequately.

Table 7.3 Main reasons for starting a business in Nima-Maamobi, by gender (%)

Reason	Gender		Total
	Male	*Female*	
Lack of employment	26	33	29
Accumulate wealth	22	21	22
Supplement household income	10	14	12
Be independent	26	23	25
Take advantage of an opportunity	15	6	11
Other reasons	1	3	2
Sample size	*65*	*73*	*138*

Source: YEMP survey (2012).

Main challenges confronting youth businesses

Schoof (2006) summarises the challenges of youth entrepreneurship as follows: sociocultural legitimacy and acceptance of young entrepreneurs, entrepreneurship education and training, access to finance and start-up capital, rigorous administrative and regulatory frameworks, and business development and support services. For poor urban communities such as Nima-Maamobi, these challenges are likely to have a severe negative impact on youth entrepreneurial activities, given the poor supporting infrastructure and services as well as the overall high level of poverty among households and individuals. Table 7.4 shows the key challenges facing youth entrepreneurship in Nima-Maamobi. It is interesting to note that the most prominent challenge – cited by almost one-third of respondents (30%) – is poor physical infrastructure and services. This is, however, unsurprising given the pressure on infrastructure and services such as water, electricity, sanitation, waste management, drainage systems, and access roads. Writing on Nima, Songsore (2003: 21–22) provides a vivid account of the intense overcrowding and poor physical environment of the community, which have remained unchanged or even worsened to date with implications for doing business there:

> There are between four to eight inhabitants per room of 9 ft by 12 ft [about 2.7 m by 3.9 m], with between 15 to 20 rooms per compound and, therefore, an average of 80 people per compound. All day-to-day living is carried on within the compound including cooking, washing, laundry and eating. Domestic animals roam freely within the compound. Because compounds are cramped, overcrowded and dilapidated, the neighbourhood is a fire hazard. The lanes are often impassable by vehicles except on foot and even then only by picking one's way past puddles, garbage and putrescent materials. Over 100,000 people live in an area of about 423 acres [about 0.7 km^2] in total defiance of any planning norms.

Table 7.4 Key challenges affecting youth entrepreneurship in Nima-Maamobi

Key challenge	Percentage
Poor physical infrastructure and services	30
Insufficient working or operating capital	20
Non-permanent business location	20
High rental charges	13
Theft/robbery	11
Scarcity of inputs/materials	3
Poor quality of inputs/materials	2
Other	1
Sample size	*138*

Source: YEMP survey (2012).

The second challenge is insufficient working or operating capital. A number of studies cite starting capital as a key constraint to youth entrepreneurship (see Chigunta et al., 2005; Baah, 2007; Jajah, 2009; ISSER, 2012). Chigunta, Schnurr, James-Wilson, and Torres (2005) argue that, although youth entrepreneurs share many of the same problems as the general population in small business development, the former differ in terms of age and other demographic characteristics, relatively low experience, capital outlay, networks, and social status. These factors directly and indirectly affect young people's ability to access capital and other resources because they are largely perceived by formal financial institutions and other lenders as being a higher-risk group. Their inability to raise start-up capital for a business leaves them reliant on family/ relatives, friends, and other informal channels. Table 7.5 indicates the main sources of start-up capital for Nima-Maamobi youth engaged in entrepreneurial activities, by gender. By far, the primary source of start-up capital is personal savings: about 77% and 62% for male and female youth, respectively. The FGDs and key informant narratives revealed that many youth had engaged in other income-earning activities – whether as employees or self-employed – through which they had accumulated funds to start their present business. Commenting on how youth mobilise start-up capital, a female FGD participant noted that

> In our community, when a lady gets to a certain age and she feels that there is something she can do, she can start saving some little amount of money. She can be saving GH₵1 per day,[4] and since she knows what she wants to do, she will do whatever she can to start it. So some people save their money little by little to get enough to start their business.

This situation partly reflects the limited support environment in which poor youth find themselves; as such, they need to do more on their own to survive in an urban environment with limited livelihood opportunities. As a female FGD participant noted:

Table 7.5 Main source of capital for starting a business in Nima-Maamobi, by gender (%)

Source	Male	Female	Total
Own savings	77	62	70
Parents	3	17	10
Other relatives	7	7	7
Boyfriend/girlfriend	3	0	2
Spouse	3	10	7
Friends	7	0	3
Other	0	4	2
Sample size	*65*	*73*	*138*

Source: YEMP survey (2012).

If your parents cannot afford a three square meal a day how can they provide capital for you to start your business? When you start on your own, they [parents] are very happy about it as they expect that you will cater for yourself and also help them. But when you are coming from a good [rich] home they will not even allow you to do business, they will want you to go to school and when you are fortunate enough, they may even send you abroad.

However, the lack of parental or family support and young people's attempts to fend for themselves can be problematic when it exposes them to bad company or illegal activities – all in the name of securing funds and other resources to start a legitimate business. As a key informant from the East Ayawaso Sub-Metropolitan Assembly noted:

Some people don't give their children even food and other basic support. As such what do you want them [youth] to do? Because you go to some homes and parents will be telling their children sometimes as young as 10 years or 11 years that their friends are all out there hustling why can't they also do something for themselves. So the young people also go out there and engage in all manner of activities so long as these activities bring money.... So some parents don't support their children and the lack of support from the parents is a problem in this community.

Indeed, this drive to strike out on their own in the context of limited family and society support is viewed in negative terms by the elderly in Nima-Maamobi. This is because youth are largely perceived as being uninterested in education and skills, craving "quick" money and material wealth instead. Commenting on the situation of youth in the past and the present, an elderly male key informant and head of a local NGO in Nima-Maamobi noted that:

In our time schooling was the main activity of the youth. But this time around, many of the youth are engaged in this *sakawa* [Internet fraud] business. I will say this is the main preoccupation of the youth today. In our time, we did not rush for money, we were preoccupied with schooling.... Nowadays though people go to school, the dropout rate is too high not because parents don't want to pay fees, no, I don't know what is going on. You'll see a small boy like 18 years riding in a car, a car costing about GH₵12,000 and you can't imagine where he got this money from. So you see, he is not thinking of education.

For the youth of Nima-Maamobi, access to credit from formal sources is likely to be further compromised by the stigmatisation and negative characterisation of the community. Writing on Nima, Owusu et al. (2008) argue that the poor state of affairs in the community – in terms of infrastructure and services and the overall physical environment – is linked to the stigma attached to its residents. Indeed, they note that the stigmatisation of Nima is sometimes the basis for denying residents both jobs and other vital services, which in turn reinforces the negative characterisation of the community (Owusu et al., 2008: 185). This affects youth in particular, who are largely perceived as violent or criminals.

Even though a large proportion of the youth surveyed indicated personal savings as their main source of start-up capital, this differs between male and female youth. Table 7.5 reveals that, relative to young men, young women tend to receive more support from their parents and spouses in terms of start-up capital. In many Ghanaian families and households, parents are expected to provide for their adult daughters as a way of preparing them for marriage. Similarly, it is common for male partners (as household breadwinners) to provide capital and other support that will enable their spouses to engage in business as a means of supplementing family or household income. These reasons probably account for the difference in sources of start-up capital among male and female youth. While young people's ability to raise capital to start their own business may be considered refreshing, for the large proportion of youth who are unable to support themselves this raises a deeper discussion on the intergenerational transfer of poverty.

According to Eguavoen (2010), youth is understood as a social category that assembles individuals in the social transition from childhood to adulthood or as "adults without adult status"– meaning individuals who have not yet established themselves socially as adults by being economically independent and taking care of themselves and other dependents. The inability of youth to prepare for this transition is due partly to limited parental support, which can lead to a situation whereby the progression to full adulthood is reversed. This is because, as Eguavoen (2010) notes, poverty can result in adults being socially relegated to the status of youth and, as a direct consequence, denied full adult rights. This leaves them with a low social status and limited access to resources. The process then feeds into another round of intergenerational poverty as today's youth transfer their poverty status to the next generation. This situation might partly account for the presence of a large number of both poor adults and youth in communities such as Nima-Maamobi.

The third challenge, cited by one-fifth (20%) of young people in relation to their businesses in Nima-Maamobi, is the non-permanency of business locations. This relates to land prices and rents as well as harassment by city and public authorities. As Baah (2007) argues, public authorities have an ambivalent attitude towards the informal economy and its operators – in many instances pursuing a silent policy of trying to eliminate the sector by forcibly removing street vendors and other informal operators who occupy public and privates spaces, or even taxing them heavily to discourage their activities. In Ghana, the implementation of the policy of decongestion by city authorities in Accra (including Nima-Maamobi) and other major cities has affected informal economy operators. This exercise, which involves the removal of informal operators from officially unauthorised locations, has hit street vendors (mostly young people) very hard as city authorities see them as little more than a nuisance.

High rents (specifically cited as a challenge by 13% of young people) also compel business relocation. Recent government slum-upgrading projects, such as road construction, are giving Nima-Maamobi a facelift. This has attracted encroachment by commercial interests and increased the presence of super-markets, commercial banks, and high-rise residential buildings. According to Owusu et al. (2008), this creates an opportunity for people with relatively high incomes to move into the neighbourhood and displace established residents either by buying off property or contributing to higher rents. This has huge implications for the welfare and livelihoods of residents, especially for youth and their businesses. Key informant interviews and FGDs with young people indi-cate that the question of high business rents is a critical concern in the com-munity. One female FGD respondent remarked:

> Yes, it is not easy getting a shop; to raise the funds for the shop is very expen-sive. It depends on the shop and the location. If you are looking at a shop along the roadside then I think you should be looking at GH₵5,000; if you are looking for a container then you should be looking for GH₵500 and a location where you will be paying a monthly fee. That monthly fee, you will have to pay to the owner for a year before you can put your container there.

Another key informant mentioned how difficult it was for young people to find space to set up their businesses because Nima-Maamobi was so overcrowded:

> Rents are expensive. Somebody can rent out one room for GH₵1,700 a year and ask for rent advance for two or three years.... For a kiosk, maybe you negotiate with the landlord/landlady who will demand an amount of say GH₵400 in order to allow someone to put a kiosk on his/her land and later the tenant will pay rent of maybe GH₵20, GH₵30, GH₵50 a month.

High rents are certainly a burden, especially for youth setting up their own enter-prises. Consequently, many young entrepreneurs operate from rooms in their already overcrowded homes. This can delay prospective enterprises in taking off or attracting the level of commercial patronage necessary to sustain them.

Concluding remarks

An appreciable proportion of the youth (about 29%) surveyed in Nima-Maamobi run small businesses. This situation is driven by the lack of (wage) employment opportunities, especially for women, and by young people's desire to be inde-pendent and accumulate wealth. The largest proportion of young entrepreneurs in this settlement started their businesses relying primarily on their own savings with some support from family members. Trading of all forms is the most common activity, followed by small-scale production enterprises such as dress-making or tailoring, woodwork, and repair services.

Youth entrepreneurship in Nima-Maamobi, however, is constrained by poor infrastructure, insufficient working and operating capital, the difficulty of securing and retaining workshops or worksites, and high workshop rents. For youth in poor and low-income urban communities, the challenges associated with engaging in entrepreneurship are extreme, given the peculiar socioeconomic constraints of these communities. Consequently, for poor urban communities such as Nima-Maamobi, the key starting point in youth entrepreneurship development is for city authorities to make a concerted effort to improve infrastructure and other basic services by upgrading the area. Such an initiative would have a positive impact on private sector and informal economic activities, including youth businesses. However, care must be taken to ensure that slum-upgrading interventions do not have the unintended consequence of displacing households and individuals, especially youth.

While providing youth who demonstrate strong business potential with the start-up and working capital they need is a laudable step – and should be promoted by the state and complemented with NGO and private sector support – such funding regimes should offer longer-term payment options and lower interest rates than those prevailing in the open market. In addition, this strategy must be supplemented with entrepreneurship education and training as well as business assistance and support programmes. The plea for entrepreneurship education and training is based on the low level of education among many prospective entrepreneurs, particularly in poor urban communities such as Nima-Maamobi. Although many factors contribute to entrepreneurial success, one easily influenced determinant of entrepreneur outcomes is education because it enables better entrepreneurial performance (ISSER, 2012). Furthermore, for young entrepreneurs, continued business assistance in the form of training and other tangible support beyond funds is critical, given their weak financial status, limited experience, skills and knowledge, and lack of familiarity with the business world. Unless these conditions are addressed, they will ultimately lead to business discontinuity among young entrepreneurs.

Notes

1 In this chapter, we use the term "youth" as defined in Ghana's National Youth Policy, that is, anyone within the age bracket of 15–35 years.
2 Ghana's education system below standards – report. Retrieved 20 November 2014 from www.ghanaweb.com (General news of Friday, 5 September 2014).
3 ECOWAS refers to the Economic Community of West African States, a regional economic and political grouping of countries in West Africa.
4 GH₵1 = US$0.63.

References

Anokye, P. A., & Afrane, S. K. (2014). Apprenticeship training system in Ghana: Processes, institutional dynamics and challenges. *Journal of Education and Practice, 5*(7), 130–141.
Awumbila, M., Owusu, G., & Teye, J. K. (2014). *Can rural–urban migration into slums reduce poverty? Evidence from Ghana* (Migrating out of Poverty Working Paper No. 13). London: DfID.
Baah, A. Y. (2007). *Organising the informal economy: Experiences and lessons from Africa and Asia.* Accra: Ghana Trades Union Congress.

Baah-Boateng, W., & Turkson, F. B. (2005). Employment. In E. Aryeetey (Ed.), *Globalization, employment and poverty reduction: A case study of Ghana* (pp. 104–139). Accra: Institute of Statistical, Social and Economic Research.

Chigunta, F., Schnurr, J., James-Wilson, D., & Torres, V. (2005). *Being "real" about youth entrepreneurship in eastern and southern Africa: Implications for adults, institutions and sector structures* (SEED Working Paper No. 72). Geneva: International Labour Office.

Eguavoen, I. (2010). Lawbreakers and livelihood makers: Youth-specific poverty and ambiguous livelihood strategies in Africa. *Vulnerable Children and Youth Studies, 5*(3), 268–273.

Ghana National Development Planning Commission. (2010). *Medium-term national development policy framework: Ghana shared growth and development agenda (GSGDA), 2010–2013* (Vol. 1). Accra. National Development Planning Commission.

Ghana Statistical Service. (1995). *Migration research study in Ghana: Vol. 1. Internal migration*. Accra. Ghana Statistical Service.

Ghana Statistical Service. (2012). *2010 Population and housing census: Summary report of final results*. Accra. Ghana Statistical Service.

Gough, K. V. (2010). Continuity and adaptability of home-based enterprises: A longitudinal study from Accra, Ghana. *International Development Planning Review, 32*(1), 45–70.

Institute of Statistical, Social and Economic Research. (2010). *The state of the Ghanaian economy in 2009*. Accra. ISSER.

Institute of Statistical, Social and Economic Research. (2012). *The state of the Ghanaian economy in 2011*. Accra. ISSER.

Institute of Statistical, Social and Economic Research. (2013). *Ghana social development outlook 2012*. Accra. ISSER.

Jajah, S. M. (2009). *The challenges of young entrepreneurs in Ghana*. Washington, DC: Centre for International Public Enterprise. Retrieved from www.cipe.org/sites/default/files/publication-docs/The%20Challenges%20of%20Young%20Entrepreneurs%20in%20Ghana.pdf.

Kararach, G., Hanson, K. T., & Léautier, F. A. (2011). *Regional integration policies to support job creation for Africa's burgeoning youth population* (Working Paper No. 21). Harare: African Capacity Building Foundation.

Langevang, T., & Gough, K. V. (2012). Diverging pathways: Young female employment and entrepreneurship in sub-Saharan Africa. *The Geographical Journal, 178*(3), 242–252.

Langevang, T., Gough, K. V., Yankson, P., Owusu, G., & Osei, G. (2015). Bounded entrepreneurial vitality: The mixed embeddedness of female entrepreneurship. *Economic Geography*. doi: 10.1111/ecge.12092.

Owusu, G., Agyei-Mensah, S., & Lund, R. (2008). Slums of hope and slums of despair: Mobility and livelihoods in Nima, Accra. *Norwegian Journal of Geography, 62*, 180–190.

Schoof, U. (2006). *Stimulating youth entrepreneurship: Barriers and incentives to enterprise start-ups by young people* (SEED Working Paper No. 76). Geneva: International Labour Office.

Sommers, M. (2010). Urban youth in Africa. *Environment and Urbanization, 22*(2), 317–332.

Songsore, J. (2003). The urban housing crisis in Ghana capital: The state versus the people. *Ghana Social Science Journal, 2*(1), 1–31.

UN-Habitat. (2009). *Ghana: Accra urban profile*. Nairobi.

UN-Habitat. (2011). *Ghana housing profile*. Nairobi.

Concluding comments to Part II

Katherine V. Gough and Thilde Langevang

The three preceding chapters, which presented the experiences of young entrepreneurs living in low-income settlements in Lusaka, Kampala, and Accra, reveal how despite living in countries with differing urban characteristics, the types of businesses they establish and the challenges they face are remarkably similar. In these concluding comments to Part II, the typical characteristics of the businesses and their operators, the support they draw on, and the problems they have to overcome are discussed and the benefits of adopting a participatory approach when researching young people in a low-income urban setting highlighted.

In the three settlements studied, a similar number of young people – between 25 and 29% – were classified as young entrepreneurs. In line with the GEM studies presented in Part I, in Uganda and Zambia a slightly higher proportion of young entrepreneurs are male whereas in Ghana there are more female young entrepreneurs (see also Langevang, Gough, Yankson, Owusu, & Osei, 2015). The most common business type by far in all three settlements is trading, especially in foodstuffs, which due to the large number of businesses is characterised by cut-throat competition. Services are the next most common enterprise with young women specialising in hairdressing and young men in activities such as shoe mending, car washing, and auto-mechanics. In all three countries, and as has been found elsewhere in a sub-Saharan African context (Filmer & Fox, 2014), manufacturing only accounts for a very small proportion of enterprises. A few activities are specific to certain cities/settlements: in Kampala, including Bwaise, many young men operate motorbike taxi services while in the Accra study dressmaking/tailoring is an activity engaged in by one-quarter of the young entrepreneurs as Nima is a Muslim-dominated area where there is a high demand for custom-made embroidered clothes. This shows how place matters and, despite the overall similarity between the experiences of young entrepreneurs in the three settlements, can influence the types of businesses they run.

The urban young entrepreneurs are extremely resourceful in the businesses that they establish, seizing opportunities when they arise and at times turning adversity into opportunity. The concept of "bricolage" was shown to be particularly apt in this respect, showing how young people make do with limited resources. None of the young entrepreneurs had obtained a loan from a financial institution for their business rather relying on their own savings, typically generated from informal/

casual employment, sometimes with additional support from family and friends. There was a clear preference among the urban youth for becoming an entrepreneur rather than being informally employed, which is viewed as being highly exploitative, and in some cases preferable to formal public or private sector employment. The majority of the businesses are run by the entrepreneur alone, though a few employ one or more additional workers, hence each business does not generate much employment and they are not viewed as representing successful businesses by young people themselves. Despite this, young people overwhelmingly expressed a desire to expand their businesses and, as the studies showed, it is not uncommon for young people to operate multiple businesses simultaneously or switch businesses as new opportunities arise. The spatial metaphor "zigzag" (Jeffrey & Dyson, 2013; Jones, 2010) is useful to highlight how young entrepreneurs are constantly having to change their ways of working and their businesses as circumstances change in order to survive.

The path to enterprise ownership for the vast majority of young people living in low-income urban settlements is neither linear nor easy and they face multiple challenges in establishing and running their businesses. A key issue frequently mentioned by the youth is lack of capital which is linked to low profits due to the high levels of competition and the inability to invest any profits back in the business due to having to support other family members. Poor infrastructure and services also feature prominently in all three studies with flooding, garbage, and unreliable water and electricity being particular problems for business owners operating out of the settlements studied. The young entrepreneurs also feel limited by their lack of education, skills training, and business support services. The insecure and hectic environments in which the young entrepreneurs are working contribute to the insecurity of their businesses (Gough, Chigunta, & Langevang, 2015).

Very few of the young people have benefited from any government policies to promote entrepreneurship; many of them do not even know such programmes exist, and those that did expressed frustration that it was almost impossible for them to gain access to the promised resources. This confirms reports that the majority of urban youth do not benefit from policies to promote youth entrepreneurship (Honwana, 2012). Partly to fill this gap, some young people become social entrepreneurs, not only concerned to generate an income but also to help others in the process. Where young people did mention having contact with the government it was almost entirely regarded as being negative as despite working informally they are still subject to regulations, such as paying fees and licences, though they did not feel that they benefit in any way. In the case of Nima, the Ghanaian government's slum upgrading programme has resulted in the settlement becoming more attractive to outside investors resulting in some young entrepreneurs being forced out due to rising rents. Hence a policy that is intended to benefit a community can have negative consequences, especially for young entrepreneurs who are unable to compete with more established business owners. These experiences do not mean that attempts to support urban youth entrepreneurship should be dropped, but rather that is it vital that awareness of and access to such programmes are widened, and that the potential implications of other policies on young entrepreneurs are considered prior to implementation.

The three urban settlement studies have also showcased the benefits of adopting participatory methodologies when studying young entrepreneurs. Not only did this improve the quality of the data collected and generate a deeper and more contextualised understanding of entrepreneurship, but more engaging research relationships were generated and the young people who worked as assistants became empowered in the process (Gough, Langevang, & Namatovu, 2014).

References

Filmer, D., & Fox, L. (2014). *Youth employment in sub-Saharan Africa*. Washington, DC: IBRD/World Bank.

Gough, K. V. Chigunta, F., & Langevang, T. (2015). Expanding the scales and domains of insecurity: Youth employment in urban Zambia. *Environment and Planning A*. DOI: 10.1177/0308518X15613793.

Gough, K. V., Langevang, T., & Namatovu, R. (2014). Researching entrepreneurship in low-income settlements: The strengths and challenges of using participatory methods. *Environment and Urbanization, 26*(1), 297–311.

Honwana, A. (2012). *The time of youth: Work, social change and politics in Africa*. Boulder, CO and London: Kumarian Press.

Jeffrey, C., & Dyson, J. (2013). Zigzag capitalism: Youth entrepreneurship in the contemporary global South. *Geoforum, 49*, R1–R3.

Jones, J. L. (2010). "Nothing is straight in Zimbabwe": The rise of the Kukiya-kiya economy 2000–2008. *Journal of Southern African Studies, 36*(2), 285–299.

Langevang, T., Gough, K. V., Yankson, P., Owusu, G., & Osei, G. (2015). Bounded entrepreneurial vitality: The mixed embeddedness of female entrepreneurship, *Economic Geography, 9*(4): 449–473.

Part III

Youth entrepreneurship in rural areas

Introduction to Part III

Katherine V. Gough and Thilde Langevang

Within sub-Saharan Africa, just under two-thirds (63%) of the population live in rural areas (World Bank, 2015), many of whom are involved to some degree in agricultural production (Sumberg et al., 2014). It is widely reported, however, that few young people see a future for themselves in agriculture or even in rural areas. This is accredited to young people's insufficient access to information and education, limited access to land, inadequate access to financial services, difficulties accessing green jobs, poor access to markets, and limited involvement in policy dialogue (FAO, 2014). The lack of involvement of young people in agriculture is typically portrayed as a problem since it results in young people migrating out of rural areas, ageing farm populations, low agricultural productivity, high levels of youth unemployment, and the persistence of rural poverty. As Sumberg, Anyidoho, Leavy, te Lintelo, and Wellard (2012) argue, this results in two complementary perspectives being promoted: "youth in peril" and "agriculture in peril".

The combined concern for high youth unemployment and low agricultural production has resulted in a wide range of policy makers and development professionals promoting youth involvement in agriculture as the solution to both issues. Consequently, across sub-Saharan Africa a raft of policies have been introduced to try to entice young people to stay in rural areas and engage in agricultural production. Such policies include ones that specifically target getting more young people involved in agriculture, typically providing improved access to essential inputs, alongside more general agricultural sector programmes that may benefit some youth and youth employment policies that may benefit some youth working in agriculture (Sumberg et al., 2014). As te Lintelo (2012) argues, however, such policies tend to prescribe one-size-fits-all solutions that do not take account of the differing needs of rural youth and the varied ways in which they engage in agriculture.

What the "problem of youth in agriculture" approach and proposed solutions tend to overlook is that for many people of all ages living in rural areas, agriculture is not their only or even their main occupation. While the engagement of rural inhabitants in nonfarm activities has long been documented, it was the "deagrarianization of the peasantry" argument, spearheaded by Bryceson (1996), which was most successful in challenging the myth that "everyone in Africa is a farmer". Rural livelihoods are typically highly diversified with many rural residents engaging in income-generating activities such as handicraft production, trading, and service provision, as well as migrating (often seasonally) to work in other rural and urban areas. Relatively little is known, however, about the rural livelihoods of young people in sub-Saharan Africa, their views on agriculture, and which businesses they establish (Sumberg et al., 2012).

In this part of the book we turn to explore the entrepreneurial activities that young people living in rural settlements in Ghana, Uganda, and Zambia engage in. In each country, two to three rural settlements with slightly different characteristics were selected for the study. Drawing on a combination of qualitative and quantitative data collected in each rural settlement, the livelihood strategies of young people are explored. The ways in which young people combine farming, running small businesses, and migrating are highlighted, along with the importance of seeing opportunities for young entrepreneurs in rural areas as not only being synonymous with agriculture.

References

Bryceson, D. F. (1996). Deagrarianization and rural employment in sub-Saharan Africa: A sectoral perspective. *World Development, 24*(1), 97–111.

FAO. (2014). *Youth and agriculture: Key challenges and concrete solutions.* FAO, CTA, IFAD. Retrieved 1 June 2015 from www.fao.org/3/a-i3947e.pdf.

Sumberg, J., Anyidoho, N. A., Chasukwa, M., Chinsinga, B., Leavy, J., Tadele, G., Whitfield, A., & Yaro, J. (2014). *Young people, agriculture, and employment in rural Africa* (WIDER Working Paper 2014/080). Helsinki: UNU-WIDER.

Sumberg, J., Anyidoho, N. A., Leavy, J., te Lintelo, D. J. H., & Wellard, K. (2012). Introduction: The young people and agriculture "problem" in Africa. *IDS Bulletin, 43*(6), 1–8.

te Lintelo, D. J. H. (2012). Young people in African (agricultural) policy processes? What national youth policies can tell us. *IDS Bulletin, 43*(6), 90–103.

World Bank. (2015). Agriculture and rural development. Retrieved 1 June 2015 from http://data.worldbank.org/topic/agriculture-and-rural-development.

8 Mobile rural youth in northern Ghana

Combining near and distant opportunity spaces

Katherine V. Gough and Torben Birch-Thomsen

Introduction

Young people in northern Ghana are growing up in a very different environment from their southern counterparts. While the south is the locus of the major cities, industries, and most important cash crops, the north is primarily rural with an agricultural base, much of it subsistence. This distinction between the southern core and northern periphery has a long history, stemming from when the country of Ghana came into being. Under colonial rule, the north was treated as a cheap source of labour to support the development of the export sector concentrated in the coastal port towns and in the southern forest belt where most cash crop cultivation took place (Songsore, 2003). Labour migration from the north to work on cocoa farms, in mines, and in urban areas located in the south has thus long taken place (Awumbila & Ardayfio-Schandorf, 2008). Import substitution industrialisation policies, followed by structural adjustment programmes and neoliberalism, have all contributed to increasing the inequality between the north and the south. Although Ghana has now joined the ranks of lower middle-income countries, its northern part lags behind with 22.2% of the population living below the poverty line of US$1.25 per day (Amanor-Boadu, Zereyesus, & Asiedu-Dartey, 2013).

Despite Ghana having recently crossed the threshold to being predominantly urban, the primary occupation for both men and women remains agriculture, forestry, and fishing; approaching half (45.8%) of all households are defined as agricultural households (Ghana Statistical Service, 2012). Widespread reports that young people are no longer interested in agriculture, however, and are leaving the rural areas in droves have resulted in a series of policies being introduced with the aim of getting young people to stay in rural areas and engage in farming (Okali & Sumberg, 2012). This approach, however, overlooks the range of nonfarm entrepreneurial activities that take place in rural areas (Anyidoho, Leavy, & Asenso-Okyere, 2012). The aim of this chapter is to examine the livelihood strategies of young people living in three villages near the town of Bole in the Northern Region of Ghana, looking at their farming and fishing activities, nonfarm businesses in the village, and the role of migration in business establishment. Arguably, since many of the young people have already lived outside their rural communities, rather than trying to stop them from moving, their mobility should be seen as a potential source of capital and experience, which those who return can use to invest in farming or establishing an alternative business.

Rural livelihoods in northern Ghana

Agriculture forms the mainstay of the economy of northern Ghana. In Bole District, where this research took place, the Ministry of Food and Agriculture (2015) reports that over 75% of the workforce is engaged in agriculture. The primary food crops are maize, millet, sorghum, rice, groundnuts, yams, and cassava, the main cash crops are cashew, sheanuts, and mango, and the key livestock reared are cattle, sheep, goats, pigs, and guinea fowl. Most farmers practise shifting cultivation and mixed cropping, with a few adopting mono-cropping and crop rotation. As is the norm in much of Ghana, land is customarily held under the control of chiefs who are responsible for its allocation. Contrary to the situation in much of sub-Saharan Africa, land is abundant in northern Ghana. The constraints to agricultural production are rather low soil fertility, reliance on increasingly erratic rainfall, and poor market access. Consequently, in Bole District, the average farm size is reported to be only 0.8 hectares (Ministry of Food and Agriculture, 2015). In order to promote farming among young people and improve agricultural production, the Government of Ghana has introduced the Youth in Agriculture Programme. The programme's main objective is to motivate young people to value farming/food production as a commercial venture and take up farming as a lifetime vocation, thus staying in their rural areas. A key component is the block farm initiative whereby land is ploughed and shared in blocks among young farmers who are provided with tractor services, subsidised inputs, and interest-free credit while under the supervision of ministry staff (Ministry of Food and Agriculture, 2015).

While agriculture remains a vital aspect of rural livelihoods, nonfarm economic activities have also become increasingly important. In a study conducted in the Upper West Region, Dary and Kuunibe (2012) find that 83% of individuals were engaged in nonfarm economic activities. Yaro (2006), studying rural communities in the Upper East Region, notes that 70% were engaged in nonfarm activities, especially during the dry season from November to March. These activities tend to be gendered: key male activities include blacksmithing, masonry, carpentry, grinding-mill operation, and mechanical repair works; female activities include brewing *pito* (local beer), sheabutter processing, food vending, and charcoal/fuel wood production (Dary & Kuunibe, 2012). Other activities such as retailing manufactured goods are generally gender-blind (Yaro, 2006). This does not mean, however, that rural residents are moving away from agriculture; rather, most combine farm and nonfarm activities (Dary & Kuunibe, 2012; Yaro, 2006).

Migration has long formed a key aspect of livelihoods in northern Ghana. Already in pre-colonial times, people migrated relatively short distances in search of fertile land and rich hunting grounds, and to escape from slave raiders (van der Geest, 2010). Under colonialism and subsequently, most migration has been north–south and primarily adult male, long-term, and long distance (Awumbila, Manuh, Quartey, Bosiakoh, & Tagoe, 2011). While the dominant trend today remains north–south, increasingly migrants consist of young people, in particular young women, who are moving independently to the cities of Accra and Kumasi to engage in menial jobs (Awumbila & Ardayfio-Schandorf, 2008). Much of this migration is seasonal, temporary, or circular, and some of those migrating are young teenagers. Since the discovery of gold in the north of Ghana, small-scale mining has also become an important rural nonfarm activity for many young people and is the motivation behind much temporary and relatively short-distance migration (Hilson, Amankwah, & Ofori-Sarpong, 2013). Consequently, young people in northern Ghana are engaging in economic activities both in situ and elsewhere.

The concepts of "near" and "distant opportunity spaces" proposed by Sumberg et al. (2014), who draw on Painter, Sumberg, and Price (1994), are useful here. Sumberg et al. (2012, 2014) define an opportunity space as the spatial and temporal distribution of work options that young people may exploit as they attempt to establish an independent life. "Near opportunity spaces" are those that are available in the vicinity of where they are living. For rural youth, these are structured around two key sets of factors. The first includes the characteristics of the rural location, in particular the availability and quality of natural resources, and the proximity and accessibility of markets – all of which affect the viability of economic activities. The second – norms and expectations – are influenced by characteristics of social difference (such as gender, age, class, ethnicity, etc.) as well as by social relations and networks, all of which help frame which activities are considered appropriate for whom (Sumberg et al., 2014). "Distant opportunity spaces" include other rural and urban areas, both national and international, which require relocating for a period of time. Thus, as Sumberg et al. (2014: 10) claim, "a willingness and ability to travel and live away from home are necessary in order to exploit the distant opportunity space". This ability to seize opportunities can be claimed to be an essential aspect of being entrepreneurial. In this chapter, we explore the ways in which near and distant opportunity spaces can be mutually reinforcing.

Methodology

The fieldwork on which this chapter is based was conducted in Bole District, one of 20 districts in the Northern Region, the capital of which is the town of Bole (see Figure 8.1). The district has a population of almost 62,000 people and a population density of just 11.9 inhabitants per km^2 (Ghana Statistical Service, 2012). With the help of a local NGO "Jacksally", three rural settlements near Bole – but with different locations and slightly different characteristics – were selected for the fieldwork. Mankuma is located on the main road going north from Bole towards Wa, 17 km from Bole; according to the assemblyman, it has a

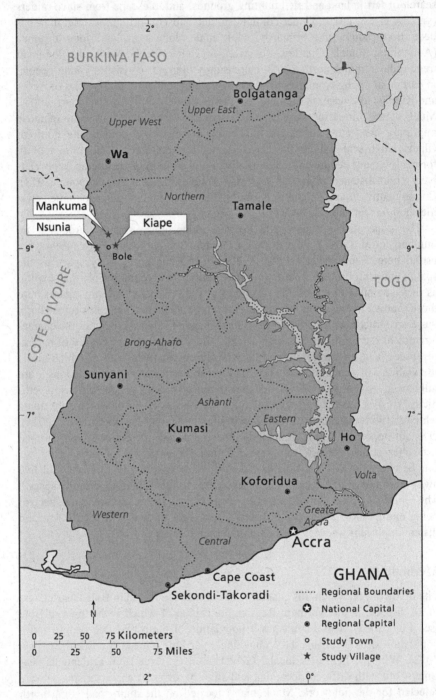

Figure 8.1 Map of Ghana showing location of study settlements (source: Kent Pørksen).

population of about 500 inhabitants who live in 58 houses. Kiape is located 10 km east of Bole along a dirt track. Nsunia is located 20 km west of Bole on the banks of the Black Volta River, which forms the border with the Ivory Coast and is home to around 840 people living in 83 houses; it is primarily a settler community of Ewes from the east of Ghana who are traditionally fishermen. Conducting the study in these three villages helped investigate the ways in which place affects the opportunity spaces of young people.

In each village, a key leading figure was approached first to explain the purpose of the research. Once permission was obtained to proceed, this person was interviewed, either alone or together with other key inhabitants, to find out the history of the settlement and hear their views on the youth situation. A focus group discussion was then held in each village with a group of 10–12 male and female youth to hear their opinions on the livelihood strategies of young people. These group discussions lasted about an hour, during which detailed notes were taken. A number of young people were then selected for individual interviews in the three villages and in Bole (19 in total). Some key government and non-government officials based in Bole were also interviewed. All the interviews were conducted by one of the authors, together with the aid of a research assistant who understood the local languages. Detailed notes were taken during the day and were written up in the evening.

A questionnaire survey was subsequently carried out in the three rural settlements, administered by Jacksally. A total of 105 randomly selected households were interviewed (35 in each village) using a similar questionnaire to the one used in rural Uganda and Zambia. Information was collected on the structure of the household, key livelihood activities, migration, land use, and agricultural practices. This chapter, however, draws primarily on the qualitative data and focuses on young entrepreneurs in the three rural villages and the ways in which they draw on near and distant opportunity spaces.

Livelihood activities of rural youth

Young people in the three study villages engage in a range of livelihood activities. These are divided here into three sections: farming and fishing, establishing a nonfarm business in the village, and migrating to work elsewhere. As the cases of specific young people illustrate, it is important to recognise that many youth combine all three strategies at various points in time.

Farming and fishing in the village

In all three villages, farming is a key livelihood activity. The main cash crops in Mankuma are cassava and yam, the former stimulated by the presence of a cassava processing plant. In addition, maize, sorghum, millet, groundnuts, and beans are grown mainly for subsistence. Shifting cultivation is practised as there is no shortage of land and it is not very fertile. One plot is typically cultivated for three years before leaving it fallow for 8–10 years. According to the assemblyman for Mankuma, farms had been as large as 20 acres in the past but nowadays 5–6 acres is considered a large farm, the most common size being around three acres. He attributed this reduction in farm size to the lower availability of labour as children now attend school and because young people are leaving to go to the mining areas rather than work on the farm. Most farms are located within a radius of one mile of the village due to the reduced demand for land. Many inhabitants have goats, sheep, and/or poultry but only a few own cattle as the land has been overgrazed. The young people claimed they are interested in farming but, due to lack of capital, are unable to engage in improved practices that would make farming more attractive for them. Almost all of them engage in farming, however, either as a primary or supplementary livelihood activity.

Similar to Mankuma, in Kiape the main cash crops grown are cassava and yam, though groundnuts are also reportedly important. Millet, maize, beans, and sorghum are also grown, mainly on a subsistence basis. Land is plentiful but tends to be located farther away from the village where it is more fertile. More of a problem is the lack of labour and expensive fertiliser. The opinion leader in Kiape explained that, although young people today are interested in farming, they are less keen than in the past because mining has become more attractive. Young people, he claimed, see their parents and grandparents who are farmers and have nothing to show for it, so they think that agriculture does not pay as it brings little reward for hard labour. Despite this, most young people living in Kiape do participate in farming.

Nsunia differs from the other two settlements in that fishing is a key livelihood activity as well as farming. Men of all ages engage in catching fish, while smoking fish and selling the fresh and smoked fish is done by women. Fishing is considered more attractive by young people than farming as it is a tradition handed down by their grandparents and because they can make a more regular income from fishing, earning money on a daily basis compared to having long periods with no income as a farmer. The fish population in the river has fallen over the years, however, and people now use finer nets – the mesh holes have decreased from 7–8 inches to 2.5–5 inches – in order to catch smaller fish, thus reducing the sustainability. Farming is still important, however, with cassava being the main crop. Yam, millet, sorghum, and maize are also widely grown and a couple of farmers cultivate cashew nuts. Most farms are around two acres in size, going up to five acres. As in the other villages, there is plenty of land but the fields tend to be located farther away from the village due to the low fertility of the immediately surrounding land. Young people still engage in farming but are much more enthusiastic about fishing as an economic activity.

According to an agricultural extension officer based in Bole, the block farm initiative has not been as popular in the area as the government expected. Although fertile land is available, the area is fairly forested and it is difficult to de-stump the land to ready it for ploughing. Consequently, most farms are small, whereas the programme was aimed at creating larger farms in order to benefit from economies of scale. In Mankuna and Kiape, only a few farmers had cleared relatively small areas of land for the initiative; in Nsunia, no one we interviewed seemed aware of it nor had they heard of the Youth in Agriculture Programme component directed at fisheries. Thus, although there are near opportunity spaces for young people to establish farming or fishing businesses, which the government seems keen to support, their viability seems to be limited by the difficulties of introducing improved practices and the possibilities of engaging in more lucrative activities.

Establishing a nonfarm business in the village

Young people have set up a range of businesses in their village (Figure 8.2). Generally, young women and men work as traders, while young men also work as masons, carpenters, butchers, and blacksmiths, and young women as hairdressers and seamstresses. It is often young people who spot new business opportunities in the villages. In Mankuma, all five shops that have been set up in the last three years are run by youth, prior to which the village only had tabletop trading done mainly by adult women. One such young person is Mohammed, an 18-year-old shop owner who was born and grew up in Mankuma. He managed to complete junior high school but could not afford to continue onto senior high school.

Like many young men, Mohammed tried his luck in mining but returned after a month as he only "got small money" and did not like life in the mining

Figure 8.2 Shop run by young man in Mankuma (photo: Torben Birch-Thomsen).

settlement. He started to run one of the first shops in Mankuma with financial support (GH₵500[1]) from his father who has a successful mango and cashew nut plantation. Mohammed's fastest-selling products are sugar, soap, and cigarettes, but he also stocks other items, including biscuits, confectionery, stationery, and plastic bags. He buys most of his stock in Wa (110 km away), mainly frequenting three stores where he is building up a relationship with the owners. As Mohammed does not have much capital he cannot buy in bulk and hence makes the trip to Wa 10 times within a four-month period, always travelling on public transport. When necessary, he supplements his stock with purchases from Bole, a journey he can make on his motorbike. While the shop is his responsibility, Mohammed is sometimes helped out by his younger brother.

Mohammed is clearly an entrepreneurial young man, always on the lookout for new opportunities. His plan is to gradually expand the shop, buying a few items to try them out before purchasing larger quantities if they sell. Once he has built up enough capital he is hoping to travel to Kumasi (340 km away) to buy stock in bulk as it is cheaper there, which would increase his profit margin. Not only is Mohammed gradually adding more stock to his shop but he has also bought a TV to entertain customers in the afternoons and evenings (made possible by the arrival of electricity in the village a few months earlier). Mohammed lets people watch TV free of charge in the hope that they will buy some items from his shop – a strategy that seems to be working. His business is doing well and he is reinvesting in it as well as saving some of his profits in the bank. Mohammed attributes his business acumen to reading books as well as some basic knowledge learned in school. He intends to stay in Mankuma and expand his business there, which he thinks is his best opportunity to make a living.

Fatima (aged 30) runs one of the other shops in Mankuma. She is not originally from Mankuma but moved there 12 years ago after marriage. Her main business is trading in foodstuffs: she goes to the surrounding villages and buys local agricultural produce, which she then sells in the market in Mankuma. To supplement this income, Fatima started a small shop seven months ago with money she had managed to save and some additional financial support from her husband, which enabled her to rent a small shop. Having studied up to primary level 6, she has some basic numeracy and literacy skills that she uses in her business. Her stock is similar to Mohammed's: the fastest-selling items are cigarettes, soap, and biscuits or sweets, though she is slowly adding goods such as condensed milk. Fatima buys on a biweekly basis in Wa where the prices are lower than in Bole. She travels there and back on local buses and always purchases from the same storekeeper with whom she has built up a strong relationship. Her children help run the shop in her absence, which enables her to operate her two businesses simultaneously. This shows how, even in a small village context, entrepreneurial young people can run multiple businesses, adding new ones when the opportunity arises. Fatima would like to have her own premises rather than rent, but is finding it difficult to save in order to build a shop – despite belonging to a *susu* savings group[2] – since she has her children's school fees to pay. She claims, however, that her business is slowly expanding, which she expects to continue given that competition in the village is relatively limited.

Aside from farming, businesses in Kiape are limited, partly restricted by the lack of electricity and unpaved roads. At the time of the fieldwork, there was just one shop run by a young man, Marshall (aged 20), and his older brother. They had perceived an opportunity to set up a small provisions store in the village two years previously, which they established with capital saved from their farming activities, thus showing how money earned in agriculture can stimulate other economic activities. Their fastest-selling products are sugar, rice, and soap, but they also sell other food and nonfood items, including biscuits, canned fish, batteries, and shaving blades. Recently, they expanded their stock of goods to include basic stationery (pens, pencils, and notebooks). They buy their stock from either Bole or Wa – sometimes paying cash – but as they build up trust with certain storekeepers they are also able to obtain goods on credit, usually paying half the amount up front and the balance later. Sometimes, Marshall and his brother sell on credit when customers do not have the necessary cash as, being a small village, they know and trust everyone. Their business is going well, partly as there is no other competition within the village, and Marshall believes they could expand the shop if they had the necessary capital. Given its small size and limited stock, however, the shop does not generate sufficient profit to be a sole income-generating activity so Marshall also engages in farming. Interestingly, he is able to manage the shop despite not having had any formal education.

In Nsunia, the river not only offers scope for fishing, but also for smuggling goods from the Ivory Coast, which is on the opposite bank. Typically, young people bring cashew and sheanuts from the Ivory Coast to sell in Ghana and take roofing sheets, cement, and sugar from Ghana across the border. The river thus offers possibilities for some entrepreneurial youth to run businesses – some of which operate on the edge of the law – that are not available in the other villages (Figure 8.3). As shown for other sub-Saharan countries, in situations of economic hardship and limited income-generating possibilities, young people seize all available opportunities and justify any illegal actions by citing the lack of alternatives (see Chapter 6 for Uganda; Gough, Chigunta, & Langevang, 2015, for Zambia; Jones, 2010, for Zimbabwe). In this way, they legitimise running businesses that technically are illegal.

This shows how young people living in rural settlements are creating new near opportunity spaces. In order to do so, they generate capital and draw on experience gained from existing near opportunity spaces (farming and fishing) and/or migrate to try to exploit distant opportunity spaces (mining and casual urban employment).

Migrating

As indicated above, it is common for young people in Ghana to migrate, especially those who have grown up in the north of the country. Some of this migration is

Figure 8.3 Young men crossing Black Volta River returning from Ivory Coast (photo: Torben Birch-Thomsen).

seasonal, while other young people stay away for a couple of years or longer (Awumbila et al., 2011). The distance young people move also varies. The survey shows that it is common for young people from the study settlements to migrate to the nearby town of Bole, where there are more opportunities to be an entrepreneur.

Young people we encountered in Bole were running businesses selling and repairing motorbikes and spare parts as well as more conventional businesses such as hairdressing, tailoring, dressmaking, and provisions stores. Not only do these businesses sell a wider range of goods and offer more services than in the rural settlements, but also some businesses are large enough to take on employees and apprentices. Young people running businesses in Bole come not only from Bole and its rural hinterland, but also from other regions (primarily the Upper East and Upper West) and from abroad such as Niger.

One of the most common forms of migration from the rural settlements studied is by young women who move to the major cities to work as porters carrying loads on their heads for a negotiated fee, known locally as *kayayoo*. This has long been an ethnic occupational niche for young northern females (Agarwal et al., 1997), who usually travel to a town where they know other young people and stay with them in a specific neighbourhood, Accra and Kumasi being the most common destinations (Awumbila & Ardayfio-Schandorf, 2008). Most of the young women who participated in the focus group discussions had worked as a *kayayoo*, staying in the city for between three months and three years. They agreed that the work and poor living conditions of the city presented a hard life but maintained it was preferable to working as a household help – which they saw as their only alternative – as the pay was better. Those who had worked as a *kayayoo* had managed to save up enough money to build a trousseau (needed before they can get married), which for many was their main incentive to leave in the first place.

A more recent trend is for young people to migrate to a small-scale mining settlement; in all three focus group discussions, almost all the young male and female participants had travelled to mining areas to work, generally staying for a number of months at a time. They had all gone to "try their luck" and while some had come back having "got something", others had "got nothing". As they explained, it was easy to migrate to a mining settlement because all they needed was the lorry fare and money for the first few days before being paid for some casual work. Similar to the *kayayoo*, young people tend to migrate to mining settlements where they already have friends who can help them settle in. The most common destination for youth from all three villages was the mining settlement of Kui, located about 80 km from Bole (see Chapter 12). According to the assemblyman for Mankuma, some of the youth who have "had luck" in the mining area have returned and bought motorbikes and built "nice" houses (i.e. with brick walls rather than mud, and sheet-iron rather than thatched roofs). This was seen as having improved the general living standards of the village, stimulated work for masons, and inspired other young people to try mining. As Hilson et al. (2013) argue, despite being a highly controversial activity with numerous associated environmental and health hazards, there is no denying that small-scale mining has brought increased wealth into the north of Ghana, stimulating other business activities.

Issah is an 18-year-old young man who was born and grew up in Nsunia. When he was interviewed, he and his brother were working in a cattle-trading business, having returned recently from the mining settlement of Kui. He told how he had completed junior high school but could not continue further because he had failed one subject. A friend who had managed to build a house in a nearby village using money earned from mining in Kui convinced Issah there was money in mining, and so he made up his mind to follow suit. Aged just 16, he travelled to Kui planning to make some money before returning to set up a fuel-selling point. When he informed his parents and his older brother of his intentions, his parents were happy that he wanted to go and "look for money" but his brother was against the idea, saying that he should not "follow bad friends". First, though, Issah had to raise some capital before he could travel. He decided to help his mother selling *kenkey* (local food) and when he had saved GH₵21, his mother gave him an additional GH₵5, his father GH₵10, and his friend GH₵5. It cost him GH₵11 to get to Kui, where he travelled with his friend in August 2009.

On arriving in Kui, Issah's friend taught him how to operate a grinding mill, which took four days to learn. He was paid GH₵10 daily for the work but then, as he put it, "I wanted to also try my luck mining." With the money earned from grinding the stones, he bought the necessary tools (a hammer, shovel, chisel, and pickaxe) for GH₵46. After digging for gold for two months, Issah had made GH₵228 from the sale of the gold. He found life difficult in Kui though. The cost of living was high as everything has to be transported down a rough dirt track making goods expensive, and he fell sick for a week after consuming poorly prepared food and contaminated drinking water. As he said, "Life was not good for me then as I was not making it." He also became very aware of the problems associated with small-scale mining when his friend was attacked by armed robbers on his way to Bole to sell his gold and had all of it stolen. When Issah returned to Nsunia, he was able to give his mother GH₵100 and his older brother GH₵200 to invest in the cattle-trading business. He also bought himself a motorbike and some clothes. In sharp contrast to mining, he now earns around GH₵5 every two weeks, working with his brother trading cattle, but believes this sum will increase with time. Issah had planned to return to Kui but heard that the situation there had become more difficult due to lack of water. Instead, his plans were to trade in small ruminants, which he would buy in the Ivory Coast and sell in the village. Issah saves GH₵3 every other week with his mother so he can build up some capital.

It is not uncommon for young people to migrate multiple times to a range of different places, especially when not every move is successful. Asana (aged 25), a married woman with four young daughters who participated in the focus group discussion in Nsunia, was keen to recount her story. She moved for the first time when she was just 15 years old, traveling to the mining town of Obuasi in the Ashanti Region to stay with her older sister. While she was hoping to learn a trade there, this did not happen as her sister expected her to work around the house in exchange for her keep. Asana subsequently moved to the city of Kumasi, also in the Ashanti Region, where she lived with another sister and worked in her shop. She married and had two children during this period.

Following a dispute with her husband, she left him and her mother provided her with funds to travel across the border to the Ivory Coast where a friend helped her find work as a maid. Her mother looked after the children in Nsunia but as Asana had to work very hard for low pay, she returned to the village after just over a year. Despite all her moves and attempts to improve her life, Asana had not been able to learn a trade or save any capital and was back living with her mother in Nsunia, helping out on the family's farm.

These cases show how young people exploit both near and distant opportunity spaces, moving between the two as new perceived opportunities arise and as others close down when perceived as being too risky or unrewarding to pursue. They also illustrate how mobility is commonplace among young people living in rural settlements in northern Ghana. Contrary to the common discourse, however, some youth do return to their villages and use the money they have earned elsewhere to set up a business.

Conclusion

Despite growing up in a part of Ghana characterised as predominantly poor and offering limited opportunities, some young people in rural northern Ghana are entrepreneurs. It is primarily young people who are spotting new opportunities and are often the first to establish a nonfarm business in rural settlements where rising disposable incomes and increasing demand for goods and services are creating a demand where previously there was none (Yaro, 2006). Some of the capital used to establish these businesses comes from in situ farming activities, while in other instances, young people raise the capital by working elsewhere, typically in mining or menial jobs in the southern cities. This highlights the symbiotic relationship between income earned in near and distant places, and from agriculture, mining, trade, and services. Some young people succeed in investing their capital in new businesses in their villages, and many simultaneously engage in a number of activities, either combining farming with a nonfarm business or adding a new business to an existing one.

While the opportunities for young people in rural settlements are broadly similar, some subtle differences emerge that can affect aspiring young entrepreneurs. As the differences between Mankuma and Kiape highlight, the availability of electricity stimulates businesses and a paved road facilitates the transport of goods. In Nsunia, which is on the banks of a river, fishing has become a key activity, while its location on the border of the Ivory Coast has resulted in some young people seizing the opportunity to make a living smuggling goods in both directions across the border. This reflects the need for an in-depth understanding of the opportunities available to rural young people in their particular settings and how these differ by location and available infrastructure.

As many of these young people, both male and female, have already lived outside their rural communities, it is essential to accept youth migration as the norm rather than as a process to be prevented (see also Olwig & Gough, 2013). As this chapter has shown, young people's mobility and attempts to find income-generating activities in distant opportunity spaces – while not always successful – can result in new capital, skills, experiences, information, and networks, which are then used to establish a farm or nonfarm business on returning to the village. Migration can thus contribute to youth continuing to engage in agriculture (van der Geest, 2010). Furthermore, in a northern Ghana context, engaging in small-scale mining can help improve livelihoods, hence, as Hilson et al. (2013) have argued, it should be regulated and supported, not demonised. Consequently, while policies to improve possibilities for youth to remain in agriculture can be beneficial, this should not be the only approach and it should not be assumed that young people will be full-time farmers – few of their parents or even grandparents were and it is unrealistic to expect youth to be. This chapter thus concurs with the findings of Sumberg et al. (2014) that it is essential to adopt a longer-term view of the dynamics between young people's near and distant opportunity spaces, between rural and urban areas, and between farm and nonfarm activities.

Notes

1 At the time of the fieldwork, GH₵1 = US$0.63.
2 *Susu* groups are savings groups, where group members arrange for collection and payment.

References

Agarwal, S., Attah, M., Apt, N., Grieco, M., Kwakye, E. A., & Turner, J. (1997). Bearing the weight: The *kayayoo*, Ghana's working girl child. *International Social Work, 40*, 245–263.

Amanor-Boadu, V., Zereyesus, Y., & Asiedu-Dartey, J. (2013). *A district-level analysis of the prevalence of poverty in northern Ghana* (METSS-Ghana Research and Issue Paper No. 01-2013). Washington, DC: USAID.

Anyidoho, N. A., Leavy, J., & Asenso-Okyere, K. (2012). Perceptions and aspirations: A case study of young people in Ghana's cocoa sector. *IDS Bulletin, 43*(6), 20–32.

Awumbila, M., & Ardayfio-Schandorf, E. (2008). Gendered poverty, migration and livelihood strategies of female porters in Accra, Ghana. *Norwegian Journal of Geography, 62*, 171–179.

Awumbila, M., Manuh, T., Quartey, P., Bosiakoh, T. A., & Tagoe, C. A. (2011). *Migration and mobility in Ghana: Trends, issues and emerging research gaps.* Accra: Woeli Publishing Services.

Dary, S. K., & Kuunibe, N. (2012). Participation in rural nonfarm economic activities in Ghana. *American International Journal of Contemporary Research, 2*(8), 154–161.

Ghana Statistical Service. (2012). *2010 Population and housing census: Summary report of final results.* Accra: Ghana Statistical Service.

Gough, K. V., Chigunta, F., & Langevang, T. (2015). Expanding the scales and domains of insecurity: Youth employment in urban Zambia. *Environment and Planning A.* DOI: 10.1177/0308518X15613793.

Hilson, G., Amankwah, R., & Ofori-Sarpong, G. (2013). Going for gold: Transitional livelihoods in northern Ghana. *Journal of Modern African Studies, 51*(1), 109–137.

Jones, J. L. (2010). "Nothing is straight in Zimbabwe": The rise of the *Kukiya-kiya* economy 2000–2008. *Journal of Southern African Studies, 36*(2), 285–289.

Ministry of Food and Agriculture. (2015). *Youth in agriculture* [Webpage]. Retrieved 1 June 2015 from http://mofa.gov.gh/site/?page_id=1173.

Okali, C., & Sumberg, J. (2012). Quick money and power: Tomatoes and livelihood building in rural Brong Ahafo, Ghana. *IDS Bulletin, 43*(6), 44–57.

Olwig, M., & Gough, K. V. (2013). Basket weaving and social weaving: Young Ghana-ian artisans' mobilization of resources through mobility in times of climate change. *Geoforum, 45*, 168–177.

Painter, T., Sumberg, J., & Price, T. (1994). Your *terroir* and my "action space": Implica-tions of differentiation, mobility and diversification for the *approche terroir* in Sahe-lian West Africa. *Africa, 64*(4), 447–464.

Songsore, J. (2003). *Regional development in Ghana: The theory and the reality.* Accra: Woeli Publishing Services.

Sumberg, J., Anyidoho, N. A., Chasukwa, M., Chinsinga, B., Leavy, J., Tadele, G., Whitfield, S., & Yaro, J. (2014). *Young people, agriculture, and employment in rural Africa* (Working Paper No. 2014/080). Helsinki: UNU-WIDER.

Sumberg, J., Anyidoho, N. A., Leavy, J., te Lintelo, D. J. H., & Wellard, K. (2012). Intro-duction: The young people and agriculture "problem" in Africa. *IDS Bulletin, 43*(6), 1–8.

Tsikata, D., & Yaro, J. A. (2014). When a good business model is not enough: Land transactions and gendered livelihood prospects in rural Ghana. *Feminist Economics, 20*(1), 201–226.

van der Geest, K. (2010). Local perceptions of migration from northwest Ghana. *Africa, 80*(4), 595–619.

Yaro, J. A. (2006). Is deagrarianisation real? A study of livelihood activities in rural northern Ghana. *Journal of Modern African Studies, 44*, 125–156.

9 Rural youth entrepreneurship in eastern Uganda

Søren Bech Pilgaard Kristensen,
Rebecca Namatovu, and Samuel Dawa

Introduction

Agriculture is the main economic activity in Uganda employing two-thirds of the population and generating 25% of the gross domestic product in 2014 (Uganda Bureau of Statistics [UBOS], 2012, 2015). Despite its current significance, the sector is projected to become less important both in Uganda and in sub-Saharan Africa in general, such that a growing share of rural incomes will come from nonagricultural activities instead (Christiaensen & Todo, 2013; Davis et al., 2010; Haggblade, Hazell, & Reardon, 2009; Winters et al., 2008). Although Uganda's rural youth have traditionally engaged in agricultural activities – learning the required skills from a young age and obtaining land from their parents – the recent literature suggests that young people are increasingly turning away from agriculture (Rietveld, Mpiira, & Staver, 2012; White, 2012). This has generated the concern among policy makers that young people will leave the rural areas for urban settlements, adding to the mass of unemployed urban youth and reducing the labour available in rural areas and, in particular, in the agriculture sector.

This chapter aims to improve our understanding of the dynamics underlying youth employment and entrepreneurship strategies in rural Uganda. It does so through a case study of young people's livelihood strategies and entrepreneurial activities in two villages in eastern Uganda. The activities and sectors in which these young people are engaged, and the challenges and opportunities for rural entrepreneurship, are analysed highlighting the similarities and differences between the two villages and between the economic activities of the old and the young. The chapter indicates that access to agricultural land is a key factor for business establishment and the lack of farmland is, therefore, one of the drivers of rural–urban migration in Uganda. Furthermore, the results show that pluri-active youth, who combine agricultural activities with nonfarm activities, seem more resilient to the prevailing challenges and uncertainties than those who specialise in a single activity.

Rural livelihoods in Uganda

An estimated 85% of the population in Uganda live in rural areas and two-thirds, predominantly smallholder farmers, depend directly on agricultural production for their livelihoods (MAAIF, 2010). Vacant land was traditionally allocated among people by the tribal authority but local administrations are increasingly involved in the legal approval of land ownership. Land that is already under cultivation is passed on within a family from one generation to the next. High population growth rates in recent decades, along with increased demand for agricultural land for large-scale commercial farming, mineral exploitation, and urban growth have reduced the land available in many parts of Uganda. The average farm size in 2008/2009 was estimated at 1.1 ha (UBOS, 2010). Many households live off subsistence agriculture and 70% of the area is used to produce locally consumed food crops (Ronner & Giller, 2012). Of these, cassava (15%), sweet potato (15%), and cooking-variety banana (14%) are the most important, followed by coffee (12%), maize, and beans. Cassava and sweet potato are regarded as food security crops. The major cereal crops are maize, finger millet, sorghum, rice, pearl millet, and wheat (for bread in urban areas). Traditional cash crops include coffee, cotton, and tea and, to some extent, tobacco produced both on estates and under out-grower arrangements on smallholder farms. Eastern Uganda's favourable climate allows the production of vegetables and fruits (citrus and pineapple production) as well as sugarcane.

While agricultural activities have traditionally generated the bulk of rural income, nonfarm income is becoming increasingly important for people living in rural areas in order to supplement and diversify their livelihoods or to replace farm income. Rural nonfarm income includes earned and unearned income received by rural people from the urban economy – through temporary migration, remittances, welfare funds, pensions, and interest – and from the rural nonfarm economy through activities such as manufacturing, processing, and construction (World Bank, 2014). The estimated share of the nonfarm sector in rural employment in the global South varies from 20 to 50%. In terms of income, Ellis (2000) states that a range of 30–50% is common in sub-Saharan Africa. The rural nonfarm sector is closely related to agriculture and the farm and nonfarm economy may be linked directly through production activities or indirectly through incomes or by investment (Reardon, Stamoulis, & Berdegue, 1998).

Several studies on rural livelihoods in Uganda have reported an increase in household participation in rural nonfarm activities in recent decades, in line with the general trend in sub-Saharan Africa. In a study of three Ugandan districts, Ellis, Ssewanyana, Kebede, and Eddie (2006) found that household levels of engagement in nonfarm entrepreneurial activities increased from 57 to 65% between 2001 and 2005 while salaried employment declined; most of these businesses were informal micro-enterprises, which did not seem to provide a strong basis for future rural nonfarm growth. In another study on Uganda, Smith, Gordon, Meadows, and Zwick (2001) found similar growth in rural nonfarm activities in two villages; they noted a sharp gender division between activities and also found that wealthier households are more likely to engage in the most profitable rural nonfarm activities, which require relatively more start-up capital. Both studies concluded that the increase in rural nonfarm activities is not necessarily pro-poor and has even widened economic differences.

This study explores the relative importance of agricultural and rural nonfarm activities for youth entrepreneurial strategies in an agricultural area where the relative proximity to urban areas offers a range of opportunities for both types of economic activities. Both agricultural and nonfarm activities may be considered entrepreneurial if they take the shape of a commercial enterprise started by an individual as an income-generating activity and which generates (at least part-time) employment for at least one person (Namatovu, Balunywa, Kyejjusa & Dawa, 2011). Specialised cash crop production (e.g. of sugarcane or pineapple) is an example of an entrepreneurial agricultural activity whereas subsistence production is not, based on the degree of commercial orientation. In some cases, surplus production originally intended for subsistence use might be sold, which makes the definition open to some uncertainty.

The study was undertaken in the villages of Butamira (located in Jinja district in the Eastern Region) and Kangulumira (located in Kayunga district in the Central Region) (Figure 9.1). Both villages are in eastern Uganda, approximately

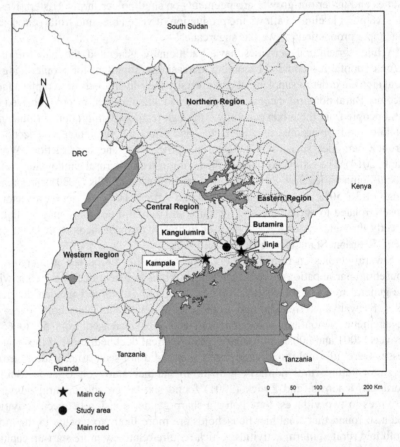

Figure 9.1 Map of Uganda showing location of study areas (source: Søren Bech Pilgaard Kristensen).

50 and 75 km east of the capital Kampala, respectively, and 20–25 km north of Jinja, the largest town in eastern Uganda with a population of about 92,000 (UBOS, 2012). Both villages are also sub-county headquarters and trading centres located on main roads and have recently been connected to the main electricity grid. They provide local services to smaller villages in the vicinity and, therefore, play an important role as local centres for public service provision, manufacturing, and trade, all of which provide a favourable backdrop for enterprise development. Since the area is fertile, agricultural production is the most important economic sector; most households grow cash crops in addition to subsistence farming. The two villages specialise in different cash crops: Butamira borders the largest sugar factory in Uganda, which has created favourable conditions for sugarcane contract farming, whereas Kangulumira is renowned for pineapple production. Maize, when produced in surplus, is another cash crop for households in both villages. The population is estimated at 7,000 in Butamira and 4,000 persons in Kangulumira (Kristensen & Birch-Thomsen, 2013).

Methodology

This chapter draws on data from two studies, both of which were conducted in Butamira and Kangulumira. As outlined above, the villages' proximity to urban areas enabled us to investigate the range of strategies adopted by youth in places offering good opportunities to engage in nonfarm activities. One of the studies, conducted under the auspices of the Youth and Employment Project (YEMP), consisted of both quantitative and qualitative data collection. A questionnaire survey was conducted across a random sample of 94 respondents (46 in Butamira and 48 in Kangulumira); the sample comprised both young (35 years and younger) and older (over 35 years) household heads. The questions dealt with sources of livelihood, economic issues, land ownership and management, and the use of extension services. In addition, focus group discussions were conducted with a purposively selected, gender-balanced group of 5–10 respondents in each village, all of whom were young entrepreneurs engaged in a variety of business sectors. The discussions explored the following themes in depth: young people's aspirations, their perceived challenges for entrepreneurship, and their views and perspectives on youth policies and programmes. In addition, key-informant interviews were held with local leaders and farmers' representatives to discuss the opportunities and challenges for young farmers, the types of agricultural innovation, and reasons for the lack of use of government programmes. The authors conducted all the focus group discussions and key-informant interviews while the questionnaire surveys were undertaken by enumerators under the authors' supervision.

The chapter also draws on data from the Rural Youth Entrepreneurship (RYE) study, which was undertaken in 2011 in both villages (see Namatovu, Dawa, Mulira, Katongole, & Nyongesa, 2012). The study consisted of a questionnaire survey of 90 young entrepreneurs (45 in each village) and a focus group discussion with 5–10 young entrepreneurs in each village, purposively selected to represent a range of business sectors and both genders. The questionnaire survey collected data on personal characteristics (age, education, gender, etc.) and enterprise (type, size, age, etc.) as well as on the role of the enabling environment (sources of start-up capital, loan taking, education and training, social networks, and governmental and nongovernmental organisation support). Both the YEMP and RYE surveys investigated the nature of youth entrepreneurship and are used in a complementary manner here: the YEMP survey, with its focus on the household as the primary unit, contrasted youth who were involved in entrepreneurial activities with those who were not, as well as "young" and "older" households, while the RYE survey focused on youth who were engaged in some form of enterprise identifying the particular challenges facing young entrepreneurs engaged in agricultural and nonfarm activities.

Activities of rural youth

Rural youth have a range of options open to them, which can be broadly divided into four categories: farming, engaging in nonfarm activities in the village, migrating, and combining activities. Table 9.1 shows the types of economic activities in which household heads, both young and older, are engaged based on results from the YEMP survey.

Table 9.1 Engagement in income-generating activities, by age of household head (%)

Activity	Butamira		Kangulumira		
	Young household head	*Older household head*	*Young household head*	*Older household head*	*Both villages*
Cash crop production	86	92	91	89	93
Farm labour	29	4	9	5	10
Employment	5	0	0	3	5
Trading	19	4	9	8	12
Manufacturing	5	0	0	0	3
Beer brewing	0	0	0	3	1
Transport	10	4	9	3	5
Other (restaurant, video hall, salon)	5	20	0	0	5
Sample size	*21*	*25*	*11*	*37*	*94*

Source: YEMP survey (2011).

Note
The total volume of activities exceeds 100% since one third of all households are engaged in several activities.

Farming

Cash crop cultivation, predominantly of sugarcane in Butamira and pineapple in Kangulumira, constitutes the most common source of income in the two villages, involving about 90% of all respondents, although with variable and often limited profitability (Figure 9.2). With its high yields and high demand, sugarcane in particular is an attractive cash crop; however, critics claim that small farmers risk jeopardising their food security by devoting too large areas to cash crop cultivation.

For young people, access to land is central to setting up a farm business. As Table 9.1 shows, young household heads are engaged in cash crop production to a similar degree as older household heads (about 90% in both villages). They are fortunate in the sense that as household heads they have been able to acquire farmland, either by inheriting or purchasing it. Nevertheless, the focus group discussions revealed that access to land for young people in general is a major challenge. Peter, a 23-year-old hardware dealer, describes how "It is much more difficult for young persons to start farming today compared to ten years ago. I cultivate a small plot that my father lends to me, but it is not my own property." The limited access to land imposes serious limitations on young people's ability to begin or expand nonfarm enterprises as the cultivation of cash crops on borrowed or rented land often is the only way for young people to generate start-up capital for other enterprises. However, the high population density and profitability of cash crop cultivation means that

Figure 9.2 Young sugarcane farmer in Butamira (photo: Søren Bech Pilgaard Kristensen).

agricultural land has become increasingly scarce. For young households in Butamira, the average farm is only 0.4 ha large as opposed to 1.85 ha for older households, illustrating that land resources are concentrated in the hands of the older generation. In Kangulumira, farm size is more evenly distributed (the average farm size is 1.2 ha and 1.1 ha for young and older households, respectively) because pineapple cultivation, which requires a lot of expensive inputs, normally takes place on a smaller area than sugarcane production, and pressure on land is therefore less than in Butamira.

Some young people are forced to sell their small piece of land to cover emergency costs (school fees or health bills) while others sell land in order to invest in an enterprise. This can be a risky strategy: if their business collapses, owners are left with no means to fend for themselves. Ali, a 20-year-old tailor, explained that he knows

> Young men who sold off their parents' land and went to urban places to invest in a motorcycle and become a *boda-boda* rider. Some of them didn't succeed and returned to the village with nothing. But, once they had returned, they did not even have a piece of land to cultivate and now they are worse off than before.

Some young people gain employment as farm labour, either casual or seasonal labour, on corporate estates (e.g. sugarcane plantations) or for other farmers with large-scale production. More young people were employed as agricultural labourers in Butamira, where sugarcane production has created a demand for casual labour for planting and harvesting, than in Kangulumira where there were fewer such opportunities (29% versus 9%, respectively).

Nonfarm activities

Other nonfarm forms of labour, including employment in professions such as the army, the police, teaching, or the church, are also found but are rare. Trading in various locations, including market stalls, shops, on the streets, and from the home, are the most common nonfarm activities. Typical items sold include foodstuffs (Figure 9.3), clothes, bicycles and motorbikes, petrol, charcoal, pharmaceutical products, and general merchandise (Namatovu et al., 2012). Trading activities are higher in Butamira (see Table 9.1), where the local economy has been stimulated by income generated from sugarcane production, and are dominated by younger people who spot new opportunities for business and as the sector is relatively easy to enter.

Manufacturing, such as tailoring and welding, and beer brewing also take place but are less common activities. Operating a transport business in the form of a motorcycle taxi (*boda-boda*) is predominantly a youth – and exclusively male – enterprise, and is common in the two villages, given that they function as trading centres. Many of the other businesses are in fact highly gendered: 75% of the youth involved in cash crop production are men, and all those running a

Figure 9.3 Young woman selling fruits and vegetables in Kangulumira (photo: Søren Bech Pilgaard Kristensen).

restaurant and 60% of those involved in trade are women (Namatovu et al., 2012). These gender differences are similar to those reported by Smith et al. (2001) in a study of rural entrepreneurship in Uganda and reflect the cultural values prevalent in many rural sub-Saharan African settings.

Migration

The YEMP study shows that 90 persons (16% of all household members) had migrated from the two villages, with this number divided evenly between men and women. One-third had moved to Kampala, while two-thirds had moved to nearby urban areas. Young people had left their village either in pursuit of education (61%) or to look for employment (22%). Social motives (marriage and divorce) together accounted for 15% of the absent household members, all of whom were women. In contrast, women constituted only 28% of the persons who had left the village to look for work, highlighting how there is a significant gender difference regarding the motives for migration. It is likely that a certain percentage of the young people who had left to study, especially those attending a secondary boarding school, might return and resettle in their villages, whereas women who had left for marriage or due to divorce will most likely not return to their village (Kristensen & Birch-Thomsen, 2013).

Discussions with youth in Kampala who had migrated from Kangulumira underline that many maintain strong links with their home village through remittances, contributions in kind, or as labour in agriculture. The migrants are ambivalent about their plans. Some have no intention of returning and stress that, although they are struggling and apt to face hardships living in the city, they still believe they have a better chance of earning an income there as there are so few economic opportunities in Kangulumira. Other young people would like to return to the village and start their own enterprise based on the skills and resources they have acquired in the city.

Combining activities

Many young people engage in more than one enterprise at the same time to generate an income or as a risk-spreading strategy. Two-thirds of the young households in Butamira were engaged in two or more activities as opposed to one-quarter of the older households. This differs strongly from Kangulumira, where a similar proportion of old and young households (20%) engaged in several activities. The difference reflects the more vibrant economic climate in Butamira. Young men typically combine cash crop cultivation with operating a *boda-boda*, brick-making, or construction work, while young women are involved in trade in addition to farming. As John, a 24-year-old *boda-boda* driver explained: "We cannot let go of agriculture. It provides us with food and cash if we are lucky. Our other enterprises may bring money but they may also collapse and then we are left with nothing." This example shows how young people combine farming with nonfarm enterprises as a way of spreading risk and using resources (mainly their labour) more profitably. During the agricultural off-season, they put more effort into their nonagricultural enterprises and vice versa. Similar risk-spreading strategies were found in a study of nonfarm activities in Ghana (Dary & Kuunibe, 2012).

Young rural entrepreneurs: challenges and opportunities

As Table 9.2 shows, the majority of young entrepreneurs surveyed explain their choice of business sector as driven by circumstances, that is, it was all they could afford to do (36%) or knew how to do (23%). In contrast, fewer people cite business advantages, such as knowledge of customers or favourable location (27%), or experience carried over from a previous business (6%). Other studies also indicate that the motives for starting a business are often mixed and may vary over time (Langevang, Namatovu, & Dawa, 2012; Rosa, Kodithuwakku, & Balunywa, 2006).

Most entrepreneurs relied on personal funds as their main source of start-up capital (69%), either in the form of savings (57%) or funds from other businesses (12%) (Table 9.3). Funds from their families (spouse or other relatives) were the most important source of capital for 21% of the entrepreneurs. It is striking that a mere 4% said they had used funds from financial institutions, despite there

Table 9.2 Reasons for choice of business (%)

Main reason for having started this particular business	Kangulumira	Butamira	Total
It is what I could afford to do	33	38	36
It is all I knew how to do	24	22	23
I had advantages that enabled business survival	33	20	27
I was previously employed in a similar business	9	2	6
Other	0	18	9
Sample size	*45*	*45*	*90*

Source: RYE survey (2011).

Table 9.3 Main source of start-up capital (%)

Source of start-up capital	Kangulumira	Butamira	Total
Savings	60	53	57
Other business	16	9	12
Other relative	11	11	11
Spouse	4	16	10
Other	2	9	6
Formal financial institution	0	2	1
SACCO	4	0	2
MFI	2	0	1
Sample size	*45*	*45*	*90*

Source: RYE survey (2011).

being numerous funds supposedly tailored to meet the needs of small businesses in the form of saving and credit cooperatives (SACCOs) and microfinance institutions (MFIs). Focus group discussions revealed that the stringent conditions attached to formal loan arrangements with financial institutions and lack of collateral constrain young people's access to such finance, an observation also made by other studies (Namatovu et al., 2012). The physical distance to the branches and offices of formal financial institutions is an additional barrier. High interest rates and fear of defaulting on their loans, which could result in the loss of land or property, are further deterrents. As Waako, a 21-year-old retailer, explained: "The interest charged on loans is very high for us and most people fear these conditions and prefer to borrow among themselves."

Some young entrepreneurs use rotational savings groups, such as "merry-go-round" schemes in which money contributed on a monthly basis by all members is pooled and given to a different member each month. Unlike formal financial loans, these schemes impose less stringent conditions, but the funds borrowed are typically very small and not sufficient to undertake large investments. They can, however, make a real difference to entrepreneurs. For Peter, a 30-year-old mason, the funds obtained through a rotational savings organisation – to which he contributes the equivalent of US$2 a month – have enabled him to buy building materials for his enterprise. Around one-third of young entrepreneurs use rotational savings groups. These are more popular among women (39%) than men (29%), partly because women tend to face greater discrimination when approaching formal financial institutions (Namatovu et al., 2012). Group members revealed that, in addition to financial support, they have also found new customers and obtained useful business contacts through the group. Others, however, remain distrustful, having been cheated themselves or knowing of individuals who have been cheated by the leaders of such savings groups.

The businesses that young people operate are predominantly small, owner-operated enterprises either with no other employees (79%) or typically just one (13%). Thus, the wider employment effect of these enterprises in the local community is very limited, as other studies have also found (Dary & Kuunibe, 2012). These types of economic activities are often referred to as "micro-enterprises" and many are not clearly demarcated from the private life and economy of the owner (Ellis et al., 2006). Nevertheless, even occasional revenue from such businesses can be used to initiate new enterprises and should not be disregarded as only an extension of private or customary activities. Most of the businesses surveyed (72%) had been set up recently and were less than five years old.

A rural location offers both advantages and disadvantages for entrepreneurial activity. As highlighted in the focus group discussions, negative aspects include the remoteness from major markets and suppliers, which means that supplies can be irregular and expensive due to higher transport costs. In addition, the longer distance to markets frequently necessitates the use of intermediaries to buy and bring products, which adds to business costs and carries the risk of fraud and associated losses. On the positive side, the distance from urban centres protects entrepreneurs from competition, lowers certain costs such as rent and labour, and tends to result in a closer, more loyal relationship with customers. Both villages have recently been connected to the electricity grid and several respondents mentioned how this had benefited their business: they could now keep longer hours and had lower power costs (compared to using private generators), although the power supply was still unstable. The introduction of electricity has also given rise to alternative businesses such as video halls and added new aspects such as the availability of cold drinks in bars and restaurants.

Conclusion

Rural youth face many challenges and dilemmas in making a living. On the one hand, they are frequently disadvantaged compared to older generations due to the lack of land and capital, which makes it increasingly difficult to enter the agricultural sector. On the other hand, they are often not interested in agriculture, which is characterised by low and fluctuating profitability, hence many opt for nonfarm activities instead. Some young people migrate – temporarily or permanently – from their home village to urban areas, where they hope to benefit from better employment opportunities and an urban lifestyle. Others choose to stay in their village and develop their own income-generating activities despite the challenges of setting up a business and establishing a viable enterprise. Rural nonfarm activities thus account for a growing share of employment and income and are seen by many as rungs up the ladder from underemployment in low-productivity, smallholder agricultural production to self-employment in the local economy. The lack of land is a serious constraint, as cash crop production on borrowed or rented land has been one of the traditional ways to accumulate capital for investment in new enterprises.

This study of youth entrepreneurship in two villages in eastern Uganda has shown that young entrepreneurs engage in a broad range of economic activities. Virtually all the respondents surveyed were engaged in cash crop production and in addition 40% were engaged in rural nonfarm activities, ranging from trade to manufacture and transport. There is a clear gender difference in the type of businesses that youth engage in: women are predominantly engaged in trade while men are more likely to be involved in growing cash crops and transport activities. The rural location offers both economic advantages and disadvantages to businesses and the advent of electricity seems to have improved business conditions greatly. As most of the young entrepreneurs do not have any employees, however, their businesses only supplement their household earnings rather than providing wider employment opportunities. Access to finance, training, and improved infrastructure are needed to augment business development. Many respondents who engage in agricultural production alongside their enterprise have highlighted the complementary benefits of pluriactivity, combining an income-generating activity that may be driven by seasonal demand with mainly subsistence food production with occasional sales. The surplus generated from one activity can be channelled into the other, thus offering some degree of stability. Many young people, therefore, prefer pluriactivity to a more specialised mono-activity strategy. On a general level, the data indicate that, despite the diversity of rural nonfarm activities, agriculture continues to play a major role as a source of livelihood for rural youth; combining this with other economic activities provides a more resilient livelihood strategy than operating a single enterprise.

As land is becoming scarce and expensive, young people need access to finance to either rent or buy land or to use as start-up capital for nonfarm enterprises. While government programmes, such as the "Youth Livelihood Programme", are in place to assist young people financially, these do not seem to have reached many youth in rural areas. The inability to meet the strict requirements of formal financial institutions is another serious challenge that restrains rural youths' access to capital. A first step to better assist rural youths would be for the government to ensure that youth programmes are sensitive to the characteristics and needs of rural youths, and to provide a framework that allows for coordination with different development partners, such as NGOs and church groups. Second, the organisation of young people into groups for savings, production, and marketing was highlighted by several respondents as a more successful approach than targeting individuals. Thus, government support to group-based activities could improve youth involvement, as recent examples with the "Youth Opportunities Programme" under the Northern Uganda Social Action Fund have demonstrated. Third, since most rural youth activities are either agricultural or financed by cash crops, a stronger youth focus in public agricultural extension activities would help support their livelihoods, such as through youth tailored training.

References

Bennell, P. (2007). *Promoting livelihood opportunities for rural youth.* Background paper prepared for the 30th Session of IFAD's Governing Council on Rural Employment and Livelihoods. Rome: International Fund for Agricultural Development.

Christiaensen, L., & Todo, Y. (2013). *Poverty reduction during the rural–urban transformation: The role of the missing middle* (Policy Research Working Paper No. 6445). Washington, DC: World Bank.

Dary, S. K., & Kuunibe, N. (2012). Participation in rural non-farm economic activities in Ghana. *American International Journal of Contemporary Research, 2*(8), 154–161.

Davis, B., Winters, P., Carletto, G., Covarrubias, K., Quiñones, E. J., Zezza, A., Stamoulis, K., Azzarri, C., & Di Giuseppe, S. (2010). A cross-country comparison of rural income generating activities. *World Development, 38*(1), 48–63.

Ellis, F. (2000). *Rural livelihoods and diversity in developing countries.* Oxford: Oxford University Press.

Ellis, F., Ssewanyana, S., Kebede, B., & Eddie, A. (2006). *Patterns and changes in rural livelihoods in Uganda 2001–05: Findings of the LADDER 2 Project.* Kampala and Norwich: Economic Policy Research Centre and University of East Anglia, Overseas Development Group.

Haggblade, S., Hazell, P. B. R., & Reardon, T. (2009, February). *Transforming the rural nonfarm economy: Opportunities and threats in the developing world* (Issue Brief No. 58). Washington, DC: International Food Policy Research Institute.

International Youth Foundation. (2011). *Navigating challenges. Charting hope: A cross-sector situational analysis on youth in Uganda* (Vol. 1). Kampala.

Kristensen, S. B. P., & Birch-Thomsen, T. (2013). "Should I stay or should I go?" Rural youth employment in Uganda and Zambia. *International Development Planning Review, 35*(2), 175–201.

Langevang, T., Namatovu, R., & Dawa, S. (2012). Beyond necessity and opportunity entrepreneurship: Motivations and aspirations of young entrepreneurs in Uganda. *International Development Planning Review, 34*(4), 439–460.

Namatovu, R., Balunywa, W., Kyejjusa, W., & Dawa, S. (2011). *Global Entrepreneurship Monitor 2010* (Uganda Executive Report). Kampala: Makerere University Business School.

Namatovu, R., Dawa, S., Mulira, F., Katongole, C., & Nyongesa, S. (2012). *Rural youth entrepreneurs in East Africa: A view from Uganda and Kenya* (Research Report No. 32/12). Dakar: Investment Climate and Business Environment Research Fund.

Reardon, T., Stamoulis, K., & Berdegue, J. (1998). The importance and nature of rural nonfarm income in developing countries with policy implications for agriculturalist. In *The state of food and agriculture 1998*. Rome: Food and Agriculture Organization.

Rietveld, A., Mpiira, S., & Staver, C. (2012, March). *Targeting young adults/young households in Central Uganda: Where is the next generation of farmers?* Paper presented at the Young People, Farming and Food Conference, Accra, Ghana.

Ronner, E., & Giller, K. E. (2012). *Background information on agronomy, farming systems and ongoing projects on grain legumes in Uganda*. Retrieved from www.N2Africa.org.

Rosa, P. J., Kodithuwakku, S., & Balunywa, W. (2006). Entrepreneurial motivation in developing countries: What does "necessity" and "opportunity" entrepreneurship really mean? *Frontiers of Entrepreneurship Research, 26*(20), 1–13.

Smith, D. R., Gordon, A., Meadows, K., & Zwick, K. (2001). Livelihood diversification in Uganda: Patterns and determinants of change across two rural districts. *Food Policy, 26*(4), 421–435.

Uganda Bureau of Statistics (UBOS). (2010). *Uganda census of agriculture 2008/2009: Vol. 1. Summary report*. Kampala.

Uganda Bureau of Statistics (UBOS). (2012). *2012 statistical abstract*. Kampala.

Uganda Bureau of Statistics (UBOS). (2015). *Quarterly GDP*. Kampala. Retrieved 15 February 2015 from www.ubos.org/statistics/macro-economic/quarterly-gdp.

Uganda, Ministry of Agriculture, Animal Industry and Fisheries (MAAIF). (2010). *MAAIF statistical abstract 2010*. Kampala.

White, B. (2012). Agriculture and the generation problem: Rural youth, employment and the future of farming. *IDS Bulletin, 43*(6), 9–19.

Winters, P. C., Essam, T. M., Davis, B., Zezza, A., Carletto, C., & Stamoulis, K. (2008). *Patterns of rural development: A cross-country comparison using microeconomic data* (Working Paper No. 08-06). Rome: Food and Agriculture Organization, Agricultural and Development Economics Division.

World Bank. (2014, March 19). *Grow rural incomes off the farm* (Brief). Retrieved 15 September 2014 from www.worldbank.org/en/topic/agriculture/brief/grow-rural-incomes-off-the-farm.

10 Rural youth in northern Zambia
Straddling the rural–urban divide

Torben Birch-Thomsen

Introduction

The Northern Province of Zambia – one of the poorest in the country – is physically and psychologically located far from the capital of Lusaka in the south.
The province is dominated by vast areas of woodland (*miombo*) with inherently
low soil fertility and a relatively low population density. The main source of
livelihood has traditionally been a subsistence-based shifting cultivation system
known as *chitemene* (Strømgaard, 1985; Trapnell, 1943). Rural youth in the
Northern Province tend to have poor education opportunities and have generally
engaged in subsistence agriculture. As is widely reported for sub-Saharan Africa,
young people are increasingly turning their backs on agriculture, seeing it as an
occupation that is back-breaking and only fit for old people (FAO, 2014).

The aim of this chapter is to explore the livelihood strategies and aspirations
of young people living in a rural area of northern Zambia, in particular to
examine their decision to either stay in or leave their village. Their strategies are
shown to fall into three categories: farming in the village, setting up a business
in the village (combined with farming), or migrating to set up a business elsewhere. The chapter shows how, contrary to the trend in much of sub-Saharan
Africa, many young people are choosing to stay in their rural villages and engage
in farming. This is partly due to the availability of land and government programmes that have been introduced to stimulate agriculture. Increasingly,
however, young people are not relying solely on farming, but are also engaging
in nonfarm activities. Some young people are shown to be highly entrepreneurial, managing to set up and run businesses despite facing constant and changing
challenges. Whether they are based in the village or in the nearby small town,
most youth engage in livelihood strategies that straddle the rural–urban divide.

Rural livelihoods in northern Zambia

Small-scale farming dominates livelihoods in northern Zambia. Farming has
been heavily influenced by government interventions, especially during two
periods. First, during the early 1980s, national policies to support agrarianisation
led to a "maize boom" in the Northern Province (Sano, 1990). The increase in

maize production was achieved primarily by involving small-scale farmers in the market economy through the provision of subsidised inputs (e.g. seed and fertilisers). However, as part of the structural adjustment policies adopted in the late 1980s and early 1990s, these subsidies were removed, which led farmers to withdraw from the market and return to traditional crops and cultivation techniques in order to secure their livelihoods (Birch-Thomsen, 1993; Kakeya, Sugiyama, & Oyama, 2006). Second, since 2002, the Farmer Input Support Programme (FISP) (formerly the Fertiliser Support Programme) has steadily reintroduced subsidised fertiliser and hybrid maize seed to small-scale farmers who are members of cooperatives or farmers' associations. The programme provides these inputs at a subsidised rate of 75% to "viable" farmers with the capacity to grow at least 0.5 ha of maize (Ministry of Agriculture and Co-operatives, 2011; Sitali, 2011). Although this has, once again, encouraged many smallholders to practise maize cultivation on larger mono-cropped fields, large differences in accessibility mean that the programme is still in the process of being introduced to more remote villages.

Land in Zambia falls under two broad tenure categories: state land and customary land. The research for this study took place in Kasama District in the Northern Province, where almost all the land falls under the customary land category and is administered by traditional authorities. In a typical village, land is allocated to residents by the village headman. As a result of extensive land use, the traditional authorities have been open to allocating land to migrants who are attracted by the availability of land and the proximity of the urban centre. As population pressure increases, land-related conflicts are becoming more common and the local authorities have become stricter about reallocating land to "outsiders". In the last four decades, this has led to a transition in farming practices from traditional, extensive, "long-fallow" shifting cultivation to "short-fallow" and permanent agriculture. The change has been driven mainly by the increased population pressure on the land, degraded woodlands, and more importantly, by government policies (subsidies) and increasing market integration (Grogan, Birch-Thomsen, & Lyimo, 2013).

Out-migration from Kasama district – mainly of men towards the Copperbelt – was very common through the 1980s (Osei-Hwedie & Osei-Hwedie, 1992). Although recent times have seen many migrants return, population densities are still relatively low: that of Kasama District (including Kasama town) was about 16 persons/km^2 in 2000 (Central Statistical Office, 2003). Finally, the growth of Kasama town, which had a population of 74,200 in 2000, has increased its connection with the rural hinterland. The town's expanding markets, coupled with the increased availability of inputs, have generated a substantial cash income opportunity in vegetable production for many of the rural villages in the hinterland. However, the quality of access roads is highly variable, resulting in disparate levels of accessibility.

The villages of Bwacha and Ngoma were selected to represent important variations in the accessibility and characteristics of rural communities in the rural hinterland of Kasama (see Figure 10.1). The two study villages are located at

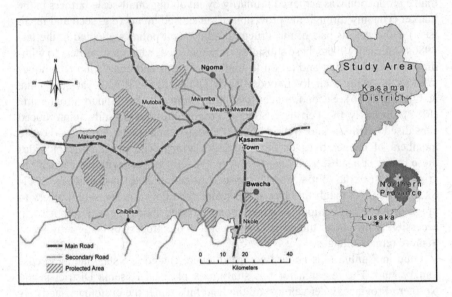

Figure 10.1 Map of Zambia and location of the study areas (source: Grogan, 2011).

approximately the same distance from the urban centre (15–20 km), but with a very different quality of access roads – Bwacha along the tarmac road south of Kasama and Ngoma along an unmaintained dirt road. As a consequence of its easier access to the urban centre, livelihoods and farming practices in Bwacha have changed, particularly during the two periods of government interventions in agriculture. In contrast, far more "traditional" practices are still found in Ngoma (including the *chitemene* long-fallow system). The generally low population densities (25 persons/km^2) and the availability of land for farming have led to in-migration in both villages, especially Bwacha, which lies near the access road to Kasama. In 2010, the estimated populations of Bwacha and Ngoma were 485 people in 85 households and 291 people in 57 households, respectively.

Methodology

This chapter draws on data from studies undertaken as part of the Youth and Employment (YEMP) project in the villages of Bwacha and Ngoma in Kasama District as well as in Kasama town, the urban and administrative centre of the Northern Province. In addition, the author conducted fieldwork in the area in 1990 as part of a PhD study. This time span of more than 20 years has enabled a longitudinal perspective on the more recent developments. Several methods – both quantitative as well as qualitative – were used to collect data. A questionnaire survey was conducted among a systematic random sample of households in both Bwacha (N=35) and Ngoma (N=35); this allowed a comparison of the livelihood strategies pursued by respondents of different age groups (youth≤35 and adults>35). Among other issues, the questions dealt with household composition, sources of livelihood, and economic activities.

Different qualitative methods were subsequently applied to obtain in-depth information on youth (aged 15–35 years) and their present livelihoods and aspirations. These included individual and focus group interviews with young people and interviews with key informants (local leaders and farmer representatives) in each rural study site (15 interviews in all). Central to these interviews were issues such as young people's aspirations, the general conditions for young farmers, and motives for (non)migration and/or mobility. Another 21 semi-structured interviews were conducted with youth in Bwacha and Kasama who had started their own business in the form of a shop or home-based enterprise; these looked at their reasons for establishing a business, the history of and future aspirations for the business activity, and the young entrepreneurs' mobility.

Young households in a rural setting

The general demographics of the surveyed households are given in Table 10.1. On average, both villages have similar orders of household size with absent household members accounting for only 2% of the population. The age distribution within the three categories is about the same in both cases, but Bwacha has a higher proportion of young household heads (37%) than Ngoma (29%) as well as of female-headed households (23% compared to 10%, respectively).

Table 10.2 gives an overview of the economic activities that constitute income sources by village and age group. The activities listed in the table fall under three main categories: agriculture, nonfarm activities, and activities requiring short-term migration or mobility. The first category dominates in terms of representation:

Table 10.1 Household composition

Composition	Bwacha		Ngoma	
		Average per household		*Average per household*
Total household members	198	5.7	180	5.1
Present household members (%)	97	5.5	98	5.0
Absent household members (%)	3	0.2	2	0.1
Age distribution				
Household members <15 (%)	46	2.6	43	2.2
Household members 15–35 (%)	36	2.0	34	1.8
Household members >35 (%)	18	1.0	22	1.1
Households with head aged 35 years or younger (%)	37		29	
Sample size	35		35	

Source: YEMP survey (2010).

Table 10.2 Household sources of income (%)

Age of household head		Bwacha (N = 35)		Ngoma (N = 35)	
		≤35	>35	≤35	>35
Agriculture	Farming	92	100	100	100
	Gardening	38	45	10	16
	Labour (farm)	15	9	20	4
Nonfarm	Labour (other)	0	5	0	8
	Tailoring	0	5	0	0
	Trading	0	5	20	16
	Hunting/fishing	15	9	0	0
	Charcoal	46	9	20	8
Short-term migration	Other	0	6	5	0
Number of activities		5	9	6	6
Sample size		*35*		*35*	

Source: YEMP survey (2010).

nearly all households, including those headed by youth, are engaged in farming upland fields ("farming" in Table 10.2, which includes crops such as maize, cassava, finger millet, and groundnuts). Gardening – the cultivation of a range of vegetables in valley bottoms, known locally as *dambos* in Chibemba – is more prevalent in Bwacha, mainly because the village is more easily accessible (see Figure 10.2). Produce is either sold directly in the market in Kasama or to intermediaries who buy directly at the farm gate (slightly more common among older households). Income from farm labour is more common among younger households and most so among the youth in Ngoma. In the latter case, farm labour typically entails weeding and cutting or pruning trees in preparation for new *chitemene* fields.

Within the category of "nonfarm" income-generating activities, the representation is more diverse, both between villages and age groups. Within "trading" and "hunting/fishing", the difference in location of the two villages is clear. Trading is more common in Ngoma, given its weaker market access, and it is an activity in which both younger and older households take part. Similarly, income from hunting/fishing in Bwacha is possible because the village has better market access, either through direct sale at the market or through traders buying in the village. There is also a higher frequency of youth selling fish, either caught in small streams or fishponds. Charcoal selling is dominated by young households; in Bwacha, nearly half the youth engage in this activity (see Figure 10.3). From the individual as well as group interviews with youth in Bwacha, it is clear that charcoal production and sale is a "male" activity and used mainly to generate capital or savings for later investment in agricultural activities (see "Strategy 1" in the next section). The category "other" involves only a few households, entailing activities that require short-term migration or mobility – in one case, a watchman's job near Kasama town.

Figure 10.2 Young farmer in his garden in Bwacha (photo: Kenneth Grogan).

Figure 10.3 Transporting charcoal on bikes to the market (photo: Kenneth Grogan).

In order to obtain a picture of the importance of these different activities, households were asked to estimate their income from each. This was difficult and likely to be very uncertain, but it was expected to indicate the relative significance of different activities. Under the FISP, more than half the households in Bwacha were receiving input and marketing support through the village cooperative (established in 2005). These households had an estimated average annual income of ZMK2,670 (US$1 = ZMK6.16), which was noticeably higher than the average income of ZMK2,200 earned by the 20% of households that were cultivating high-yield varieties. At the time of the questionnaire survey, no support was being received from the FISP and households reported earning very low incomes from traditional shifting agriculture (on average, ZMK200 per household per year). A related income difference between the two villages was within farm labour: the average income in Bwacha was twice that reported in Ngoma. In general, income levels were higher in Bwacha, which is explained primarily by its better access to the urban market. Finally, when looking at the degree of household income diversification between the two villages and age groups, some interesting differences emerge. First, for all households surveyed, the average number of activities per household in Bwacha is 2 (range = 1–5), which is slightly higher than the average of 1.5 (range = 1–3) in Ngoma. Second, young households tend, on average, to engage in more activities, mostly so in Bwacha – 2.3 activities compared to 1.7 in Ngoma.

In summary, households in Bwacha – in particular, younger ones – have diversified their income portfolios both within agricultural and alternative nonfarm activities. This has been possible through the government support provided to the agricultural sector and as a consequence of the rapidly growing Kasama market. In past decades, there has been a steady transition towards a more cash-driven economy. In addition to selling basic farm produce, close on half the households grow high-value vegetable crops sold to traders or by individuals at the market in town. In particular, producing and selling charcoal among youth is reported to have expanded since 2005 and the village headman has allocated part of the village area specifically for this purpose. Unlike Bwacha, Ngoma is less driven by a cash economy, mainly due to the lack of adequate infrastructure – the poor access road to Kasama makes transport opportunities less frequent and more costly and time consuming. Despite these unfavourable conditions, some households (of which the majority are young) have started to engage in activities linked to the growing market in Kasama, mainly related to vegetable production, trading, and charcoal production. In 2010, the village cooperative comprised only a few members but two years later this number was rapidly growing; out of 17 youth at a group meeting, 10 said they were members. This illustrates not only the increased importance of the cooperative and FISP support, but also young people's growing interest in agriculture (Kristensen & Birch-Thomsen, 2013).

Three strategies for rural youth

This section presents three main strategies for young rural households that are characteristic of the study areas.

Strategy 1: farming in the village

With the establishment of the cooperatives, fostered by subsidised input support from the government, young people in both study villages have started showing an increased interest in farming as a livelihood. This emerged clearly from the group discussions and individual interviews held. A key challenge, though, was the membership procedure. In Bwacha, young people explained that, in order to become members of the cooperative, farmers had to pay a one-off membership fee of ZMK50,000. Furthermore, to receive subsidised inputs (hybrid maize seed, fertilisers, and pesticides), at least one share worth ZMK100,000–150,000 had to be purchased. Although this represents a considerable sum for a young household, most youth respondents said they had either managed to become members or aspired to do so. Two of the most common ways of accumulating capital to invest in membership of the cooperative include the production and sale of vegetables and/or charcoal. Vegetable production, such as tomatoes, potatoes, and cabbages, takes place in the lowlands where water is available in the dry season (*dambo*) – an area not used in the traditional cultivation system. These high-value crops are either sold in the village to traders from town or transported by bicycle to the market in Kasama town.

A young man in Bwacha (30 years old) explained that he was able to harvest tomatoes and rape three times a year from his garden. In 2010, he harvested approximately eight boxes of tomatoes per harvest, receiving ZMK25,000–30,000 per box. The money saved was invested in his cooperative membership. Other youth mentioned charcoal burning and sale as their source of income for savings. For example, a 29-year-old man who had moved to the village after marriage explained how he had saved money for the membership fee and was now saving to buy a share of inputs through his charcoal business. He started the business in 2009 and has been able to produce as much as 30–40 bags per month. Unlike others who transport the charcoal to Kasama on bicycle, he only sells to traders coming to the village (at approximately ZMK10,000 per 50 kg). By 2012, he had saved enough to pay the fee and expected to be able to soon buy a share; once this is achieved, he will stop the charcoal business and concentrate on farming.

While many of the young people aspire for a future in the agriculture sector, relatively few respondents reported having benefited from government programmes (see also Izzi, 2013) and none were aware of any programmes that directly targeted youth. The agricultural intensification programmes being implemented (such as the FISP) have been criticised for imposing strict requirements on prospective beneficiaries. In general, households involved in the FISP report being better off and having a more optimistic outlook (Grogan et al., 2013). Although challenged by the difficulties of raising funds, young people in Bwacha (and later in Ngoma) saw membership of the village cooperative – and thereby access to the benefits of the FISP – as an option that would help them fulfil their aspirations for a better life in the village. Most youth seemed to want to stay in the village and, at the same time, make use of possible links to Kasama. Statements such as "food security is good here", "good land is easily accessible here", and "good communication and easy access to the town" all confirm this tendency.

Strategy 2: combining farming and small-scale business in the village

In addition to farming, most youth in these villages are diversifying their livelihoods, in some cases through the small-scale sale of produce in town or petty trade within the village, while a few have invested in starting up small-scale village shops. In Bwacha, an entrepreneurial young couple in their early 30s are running a small shop next to their house as an additional income-generating activity to augment their farm income. The shop was established in 2009 and is run by the couple and their daughter; they have no employees. It is the only (private) shop in the village except for small cooperative shops. The start-up capital was generated after a bulk harvest where the surplus maize was sold through the village cooperative (of which the young man is a member). They started by investing in a few goods for the shop: two bottles of cooking oil, some rice, sugar, sweets, and soap. Cooking oil, sugar, and rice were repacked into smaller bags from the bigger bags bought in Kasama. These goods are similar to what is sold in the cooperative shop. The idea of starting a shop was inspired by the young woman's parents, who run a small shop in Kasama, and by the increased availability of money among the villagers from selling crops through the cooperative.

The shop has been well received by the villagers and the first year was a success. The income generated was reinvested and the couple have expanded the business with new items such as batteries, dried fish (different varieties), lotion, shampoo, medicine (painkillers), and airtime for mobile phones. The dried fish was introduced a week before the interview and was a big success – "people need relish for the *nshima* [maize/cassava porridge]". The husband travels approximately 20–30 km by bicycle to buy the dried fish from fishermen living near the Chambeshi River. Most other goods are purchased through wholesale stores in Kasama. They travel to Kasama on bicycle approximately two to three times a week to buy supplies for the shop – always using cash. In addition to looking after the shop, the young woman works in the fields from 5 to 10 a.m. Thinking back on the past year-and-a-half, she thinks it has been worth the effort. They are doing well and have plans to further expand the shop – both physically and in terms of the goods on sale, which will be exclusively available here and include spare parts for bicycles, clothes, shoes, and plates.

The establishment of small shops was also observed in Ngoma in 2012 which had not been there two years previously. Three small shops had recently been set up by young men all in their mid-20s: in one case, the start-up capital had come from selling agricultural products in Kasama, and in the other two, from petty trade and charcoal production. Many of the youth interviewed in both villages mentioned specific plans for future businesses, such as buying piglets to raise for meat, chicken farming, investing in a treadle-pump to irrigate gardens in the dry season, starting a business in the village, or selling fish. This shows how some young people who wish to remain in the village develop innovative business plans.

Strategy 3: migrating to establish a business

Another livelihood strategy for young people is to migrate. In Ngoma, some young people have migrated temporarily (in the short term) for purposes of work: one young man mentioned working on a coffee farm for two months. However, for these young people, the village was still their home. Overall, in the two villages studied, there was very little permanent migration. Village leaders in Bwacha helped identify three young men who had migrated to Kasama in the last 10–15 years; one of them was located and his story of migrating to Kasama is summarised here.

At the age of 19, Geoffrey, who was born in Bwacha, decided to move with his wife and two children to Kasama. This was in 1998, four years after the first government-supported subsidy programme had ended and when earning an income from farming in the village was difficult. Using their small savings with additional support from the family, Geoffrey bought a small plot of land on the outskirts of Kasama (which has now grown beyond their plot). Initially he generated an income from buying up cassava in nearby villages and selling it at an open market near the Tazara station (the Tanzania–Zambia railway connecting Lusaka and Dar es Salaam). The surplus generated was saved in order to finance the clothing business he hoped to establish. By 2003, Geoffrey had generated enough money to rent a small shop at the Chambeshi market located in one of Kasama's fast-growing suburbs, not far from the Tazara station, south of the centre of Kasama. He started selling clothes bought in the border towns of Nakonde/Tunduma between Tanzania and Zambia, relying on the Tazara railway for transport. Over the next three years, business was good and by 2006, Geoffrey managed to join the traders' association *Twikafane* ("come together" in Chibemba). The cost of becoming a member was ZMK50,000 and, at the same time, he managed to buy the shop he had previously rented.

Geoffrey attributes his business success to his membership of the traders' association and access to its credit facility. Twice, he has received a loan from the association to invest in the shop – both to carry out renovations and purchase new stock. Although the loans carried a very high interest rate (25%) and had to be repaid within a month, he was able to repay both in time. Once, he tried to obtain a bank loan but without success. Presently, Geoffrey sells various clothing items: trousers, dresses, skirts, shirts, and bags. Apart from selling to individual buyers, he also barters goods with people who arrange clothes for sale in villages and then pays afterwards. He does not employ anyone else, but has a good friend who helps him look after the business; otherwise, he and his wife run the shop together. With the income from the shop, Geoffrey has built a new house on the plot he owns – a four-room brick house with a corrugated iron roof, furnished with a sofa, table, and chairs, and with access to electricity, radio, and television. Furthermore, the business enables him to support his family back in Bwacha with clothes and some money – approximately three times a year, adding up to about ZMK1 million over two years.

When interviewed in 2010, Geoffrey had plans to extend the shop to house more stock and did not foresee any threats to his business in the near future. However, in a follow-up interview in 2012, things had changed slightly. He complained about increased competition: there were too many similar shops in the market, including a large shop that had been opened by Chinese selling at lower prices. As a result, he had started working three days a week at a hotel to supplement his income. Apart from spending two days in the shop, Geoffrey and his wife have started to cultivate fields in Bwacha, mainly for food consumption. Despite these problems, he still has ideas for how to move forward and plans to diversify into the grocery business.

This case shows how young people who migrate to urban areas can successfully set up a business (Figure 10.4), but constantly face challenges – including increasing competition – and have to adapt their business accordingly. By maintaining close contact with their home village, young people can produce crops for their own consumption, thus reducing their expenditure on food. Producing

Figure 10.4 Small shop in Kasama run by a migrant from Bwacha (photo: Torben Birch-Thomsen).

cash crops is another way of generating income that can be invested in their urban business or kept as a security net for the future. Thus, many young people living in rural areas engage in livelihood strategies that cross the rural–urban divide.

Conclusion

Contrary to common perception in the literature on rural youth in sub-Saharan Africa, many young people around Kasama have chosen to remain in the rural areas where they see better prospects of success than if they were to migrate to an urban area. In fact, in-migration has even been recorded into the rural areas, supporting Potts' (2012) argument that Zambia is, in fact, de-urbanising. The difference between the rural youth strategies in the two villages arises primarily from differing access to land and market/job opportunities. As the chapter shows, given the opportunity, the rural youth have responded positively to favourable market conditions and external agricultural policy. Although not directed specifically at youth, the government agricultural programmes have created an alternative that enables young people to stay in the rural hinterland and make a living. There are also indications that urban living is becoming increasingly challenging, and some young people who had migrated to Kasama are returning to their rural villages to support their livelihoods through farming.

In general, most youth aspire to a future within their rural settings, pursuing additional opportunities for income in petty trade and small-scale business – this is particularly the case in rural villages with good or reasonable access to the urban centre of Kasama. This is made possible through rural–urban linkages and young people thus benefit from both livelihood options in the rural as well as urban sphere. For those who do chose to move to the urban centre, the greater demand opens up opportunities for establishing a business in one of the marketplaces in town. However, it seems to be important for this group to maintain close links to their place of origin. This "construction of a promising future" by youth and of "moving forward" despite challenging conditions, ties in well with Locke and te Lintelo's (2012) findings from elsewhere in Zambia. It is thus important that young people's desires to stay in rural areas and engage in agriculture are recognised by the Zambian government and NGOs, and that these institutions continue to adopt policies and offer support that enable the youth to do so.

References

Birch-Thomsen, T. (1993). *Effects of land use intensification: Introduction of animal traction into farming systems of different intensities* (parts I–III). PhD thesis, Geographica Hafniensia A (2), Publications – Institute of Geography, University of Copenhagen.

Central Statistical Office. (2003). *Zambia: 2000 census of population and housing.* Lusaka: Author.

FAO. (2014). *Youth and agriculture: Key challenges and concrete solutions.* FAO, CTA, IFAD. Retrieved 1 June 2015 www.fao.org/3/a-i3947e.pdf.

Grogan, K. (2011). *Land use/cover change in Kasama district, Zambia: Agricultural transition in Miombo woodlands.* Unpublished master's dissertation, University of Copenhagen, Denmark.

Grogan, K., Birch-Thomsen, T., & Lyimo, J. (2013). Transition of shifting cultivation and its impact on people's livelihoods in the Miombo woodlands of northern Zambia and south-western Tanzania. *Human Ecology, 41*(1), 77–92.

Izzi, V. (2013). Just keeping them busy? Youth employment projects as a peacebuilding tool. *International Development Planning Review, 35*(2), 103–118.

Kakeya, M., Sugiyama, Y., & Oyama, S. (2006). The *citemene* system, social levelling mechanism, and agrarian changes in the Bemba villages of northern Zambia: An overview of 23 years of "fixed-point" research. *African Study Monographs, 27*(1), 27–38.

Kristensen, S. B. P., & Birch-Thomsen, T. (2013). Should I stay or should I go? Rural youth employment in Uganda and Zambia. *International Development Planning Review, 35*(2), 175–201.

Locke, C., & te Lintelo, D. J. H. (2012). Young Zambians "waiting" for opportunities and "working towards" living well: Life-course and aspiration in youth transitions. *Journal of International Development, 24*(6), 777–794.

Ministry of Agriculture and Co-operatives. (2011). *Farmer Input Support Programme implementation manual: 2011/2012 agricultural season.* Lusaka: Ministry of Agriculture and Co-operatives.

Osei-Hwedie, K., & Osei-Hwedie, B. (1992). Reflections on Zambia's demographic profile and population policy. *Journal of Social Development in Africa, 7*(1), 87–97.

Potts, D. (2012). What do we know about urbanisation in sub-Saharan Africa and does it matter? *International Development Planning Review, 34*(1), 5–21.

Sano, H.O. (1990). *Big state, small farmers: The search for an agricultural strategy for crisis-ridden Zambia.* Copenhagen: Centre for Development Research.

Sitali, S. M. (2011). *Effect of the reduced farmer input support programme pack size on small-scale farmers' maize yield: A case of Katuba constituency (Chibombo district, Zambia).* Unpublished master's dissertation, Van Hall Larenstein University of Applied Sciences, Leeuwarden, the Netherlands.

Strømgaard, P. (1985). The infield-outfield system of shifting cultivation among the Bemba of south central Africa. *Tools and Tillage, 5*(2), 67–84.

Trapnell, C. G. (1943). *The soils, vegetation and agriculture of north-eastern Rhodesia: Report of the Ecological Survey.* Lusaka: Government Printer.

Concluding comments to Part III

Katherine V. Gough and Thilde Langevang

The preceding three chapters have shown how young people living in rural settlements in Ghana, Uganda, and Zambia are engaging in broadly similar livelihood strategies that often involve combining farming, establishing nonfarm businesses, and migrating at some point during their life course. There are some subtle differences though between the three countries and between rural settlements within each country, which will be drawn out here.

Despite the widespread belief that young people are not interested in agriculture, most youth living in rural settlements do engage in farming, though it is rarely their only occupation. For young rural Zambians in particular, farming is seen as an attractive business, while this is less the case in Ghana and Uganda. This notable difference is due to a combination of factors including the availability of land, the support of government programmes, and increasingly difficult conditions in urban areas of Zambia (Potts, 2015). In contrast, Ugandan youth find it difficult to access land, while for Ghanaian youth the possibility of earning money in small-scale mining or urban areas makes farming less attractive. Moreover, government policies to promote agriculture in Ghana and Uganda seem to have missed the mark, whereas long-standing agricultural policies in Zambia, including subsidising fertilisers and seeds, seem to have succeeded in promoting agricultural production. Across the study sites in all three countries, however, few young people reported having benefited directly from any government programmes and they were generally unaware of any programmes targeted specifically at youth. This lends credence to Sumberg et al.'s (2014) assertion that many of the current programmes introduced to promote youth agriculture do not succeed in attracting large numbers of young people to farming. The Zambian case is important, however, for highlighting how, if conditions are right, young people will engage in farming as a business, and many of the Ghanaian and Ugandan youth claimed that they would also do so if farming was more attractive. Hence it appears that it is not necessarily agriculture per se that many young people are keen to avoid but rather the specific conditions that make it unattractive as a business.

As the three chapters illustrated, however, even when rural young people do engage in farming it is rarely their only income-generating activity nor is it necessarily a business they envisage having a long-term future in. Young entrepreneurs in all three countries establish nonfarm businesses in rural settlements, often being the first to seize new opportunities as they arise. These businesses are typically in trading especially foodstuffs, service provision such as hairdressing and carpentry, and there is a limited amount of small-scale production. While both male and female youth set up small shops, most service provision is gendered according to the activity. The capital to establish these nonfarm rural businesses in some cases comes from profits generated by farming, either by the young people themselves or other family members (often parents), or in other cases from money earned elsewhere through migration. As Okali and Sumberg (2012) also found, young entrepreneurs in rural settlements do not only establish one business nor necessarily remain in the same business for lengthy periods of time. Similar to urban areas, as shown in Part II, rural young people tend to add to existing businesses, change their line of business, or set up new businesses as they perceive new opportunities, thus diversifying and running multiple businesses simultaneously.

Migration is a key aspect of many rural young people's lives in sub-Saharan Africa, often linked to perceived economic opportunities (Potts, 2015). Although frequently overlooked in the literature, as these chapters have shown, nearby small towns are important sites for migration of rural youth who perceive the opportunities to establish a business there to be greater and more lucrative than in their villages. Migration further afield, especially to large urban centres but also to other rural areas or small-scale mining settlements, is also very common among young people, though to varying extents in differing contexts. Out of the three countries studied, rural young Ghanaians are the most mobile; all of the young people interviewed in northern Ghana had spent varying periods of time living outside of their rural communities and many had plans to leave again. Young Ghanaians are known for being highly mobile, moving around as they perceive new opportunities (Langevang & Gough, 2009). In comparison, the trend in Uganda is for young people to move out of their rural settlements, especially to urban settlements, only returning for short visits, while in Zambia, many rural young people have either not left their village in the first place or have returned from urban areas where life has become very tough. These different trends emerging from the three countries are linked to both current opportunities and long-standing cultural traditions. It is important that rather than migration being portrayed as negative, as is often the case, it is seen as an integral part of young people's livelihood strategies that does not necessarily result in them abandoning rural areas and can lead to them becoming entrepreneurs.

As well as there being some key national differences between rural young entrepreneurs in the three countries studied, there are also some important differences in young people's opportunities between rural settlements located relatively close to each other. Factors such as the quality of access roads, and hence the ease of accessing markets, and the availability of electricity directly affect the ability of young people to establish businesses and the types of businesses that are feasible. As well as infrastructure, locational factors, such as being situated on the banks of a river or close to a national border, also influence the nonfarm activities that young people can engage in. The three chapters in this part thus support Sumberg et al.'s (2014: 11) claim that when considering rural youth it is imperative to take into account the "diversity of employment types ..., the diversity of rural areas and the diversity of young people". Furthermore, the studies have highlighted how blurred the divide between rural and urban is. While the young people are portrayed in these chapters as being rural youth as they are living in rural areas, most have spent some time living in urban areas and many are engaged in occupations that tend to be seen as being urban. As Tacoli and Mabala (2010), among others, have argued it is important to recognise the links between mobility and livelihood diversification and the ways this contributes to the close intertwining of rural and urban areas.

References

Langevang, T., & Gough, K. V. (2009). Survival through movement: The mobility of urban youth in Ghana. *Social and Cultural Geography, 10*(7), 741–756.

Okali, C., & Sumberg, J. (2012). Quick money and power: Tomatoes and livelihood building in rural Brong Ahafo, Ghana. *IDS Bulletin, 43*(6), 44–57.

Potts, D. (2015). Debates about African urbanisation, migration and economic growth: What can we learn from Zimbabwe and Zambia? *The Geographical Journal.* doi: 10.1111/geoj.12139.

Sumberg, J., Anyidoho, N. A., Chasukwa, M., Chinsinga, B., Leavy, J., Tadele, G., Whitfield, S., & Yaro, J. (2014). *Young people, agriculture, and employment in rural Africa* (WIDER Working Paper 2014/080). Helsinki: UNU-WIDER.

Tacoli, C., & Mabala, R. (2010). Exploring mobility and migration in the context of rural–urban linkages: Why gender and generation matter. *Environment and Urbanization, 22*(2), 389–395.

Part IV

Youth entrepreneurship in specific sectors

Introduction to Part IV

Katherine V. Gough and Thilde Langevang

The opportunities and constraints for youth entrepreneurship not only differ from country to country, and between rural and urban areas, but also between different economic sectors. The various sectors of an economy differ in relation to their structure, the type of business activities involved, the nature of the demand and supply, the resources and skills needed to start up, the policy environment surrounding the sector, and their significance and potential for growth and employment creation. Existing literature on entrepreneurship in sub-Saharan Africa has in particular highlighted the underutilised potential of farming and agroprocessing (Sumberg et al., 2014), the role of female traders (Darkwah, 2007), and common vocations for young people to enter such as hairdressing and dressmaking for young women (Langevang & Gough, 2012) and carpentry and automechanics for young men (Yankson & Owusu, 2015). In this part of the book we turn our attention to important sectors of the economy that have often been overlooked despite attracting a large number of young people in sub-Saharan Africa, namely the mobile telephony sector, small-scale mining, handicrafts, and the tourism industry.

Growth in the mobile telephony sector in sub-Saharan Africa has been remarkable in recent years, resulting in some commentators describing the mobile phone phenomenon as creating a "revolution". Apart from generating new possibilities for entrepreneurs through innovations, such as mobile banking and mobile phone applications, a large informal economy around mobile telephony has also emerged which is providing new opportunities for many young people to engage in entrepreneurship (Etzo & Collender, 2010). Like mobile telephony, small-scale gold mining has expanded dramatically in recent years linked to rising gold prices in the global market. Despite being a highly controversial livelihood activity, because of its illegal character and the severe environmental and health hazards involved, small-scale mining attracts increasing numbers of young people in sub-Saharan Africa especially from rural areas (Hilson, Amankwah, & Orfori-Spring, 2013).

The handicraft sector, on the other hand, is a long-standing livelihood activity in which the skills involved are typically passed on from generation to generation, especially by women. The sector, however, is undergoing a number of changes that are impacting on the entrepreneurship opportunities for young people (Harris, 2014). The tourism industry in sub-Saharan Africa, while not a new sector, has witnessed sustained growth in recent years due to increasing incomes among the global middle classes and a greater interest in discovering new travel destinations. Being service-based, the industry is a labour-intensive growth sector offering a growing potential for young people in terms of providing entrepreneurial opportunities and employment.

This part of the book examines these four sectors in specific country contexts: the mobile telephony sector and small-scale mining in Ghana, and handicrafts and the tourism industry in Uganda. In each case the types of business activities and nature of the work involved, the motivations of young people for entering these sectors, and the constraints and opportunities they face are discussed.

References

Darkwah, A. K. (2007). Making hay while the sun shines: Ghanaian female traders and their insertion in the global economy. In N. Gunewardena & A. Kingsolver (Eds), *The gender of globalization: Women navigating cultural and economic marginalities* (pp. 61–83). Oxford: James Currey.

Etzo, S., & Collender, G. (2010). The mobile phone "revolution" in Africa: Rhetoric or reality? *African Affairs, 109*(437), 659–668.

Harris, J. (2014). The messy reality of agglomeration economies in urban informality: Evidence from Nairobi's handicraft industry. *World Development, 61*, 102–113.

Hilson, G., Amankwah, R., & Orfori-Sarpong, G. (2013). Going for gold: Transitional livelihoods in Northern Ghana. *Journal of Modern African Studies, 51*(1), 109–137.

Langevang, T., & Gough, K. V. (2012). Diverging pathways: Young female employment and entrepreneurship in sub-Saharan Africa. *The Geographical Journal, 178*(3), 242–252.

Sumberg, J., Anyidoho, N. A., Chasukwa, M., Chinsinga, B., Leavy, J., Tadele, G., et al. (2014). Young people, agriculture, and employment in rural Africa (WIDER Working Paper 2014/080). Helsinki: UNU-WIDER.

Yankson, P. W. K., & Owusu, G. (2015). Youth entrepreneurship in auto spare parts sales and repair service in Accra, Ghana. *Developing Country Studies, 5*(4), 84–97.

11 Young entrepreneurs in the mobile telephony sector in Ghana

Robert L. Afutu-Kotey

Introduction

The mobile telephony sector in Africa has grown at a remarkable pace over the past decade and the continent continues to record the fastest growth in the number of people with mobile phone access (Aker & Mbiti, 2010; Etzo & Collender, 2010). This has led some researchers to describe the phenomenon as having created a "revolution" (Etzo & Collender, 2010: 659), although the rate of growth has not been uniform across the continent. As the subscriber base of various networks expands, a large informal economy has emerged, providing opportunities for many young people across sub-Saharan Africa to engage in self-employment and entrepreneurship (Chiumbu & Nyamanhindi, 2012; Etzo & Collender, 2010). In Ghana, this group is dominated by young people engaged in selling airtime, mobile phones, and phone accessories, or offering services recharging phone batteries and repairing mobile phone sets.

Despite the increasing research interest that has accompanied these developments, entrepreneurial activities within the mobile telephony sector have been overlooked.[1] Accordingly, this chapter explores the various types of mobile telephony businesses young people engage in. It draws on extensive fieldwork conducted in the city of Accra with young entrepreneurs working in the sector to identify which mobile telephony businesses young people establish, who these young people are, why they decide to go into such businesses, the various kinds of support available, and young people's aspirations for working in the sector. In examining these issues, the chapter seeks to assess the role of informal mobile telephony businesses as avenues for generating employment through entrepreneurship among young people.

Emergence of the mobile telephony sector and spatial concentration in Accra

Influenced by the World Bank and other international development organisations, as well as by global changes within the telecommunications sector, Ghana embarked on a programme to liberalise the sector in the early 1990s (Wallsten, 1999). The aim of these initiatives was, among others, to enable private sector participation in the provision of services in order to increase access and coverage. To this end, one strategy involved liberalising the mobile telephony subsector and introducing changes to the regulatory framework of the telecom sector as a whole (Frempong, Esselar, & Stork, 2005). Following the liberalisation of the Ghanaian economy, licences were issued to four mobile telephony companies: Millicom Ghana Limited (a subsidiary of Millicom South Africa, currently operating as Tigo), Kasapa Ghana Limited (a joint venture between Kludjeson International and Hutchinson Whampoa of Hong Kong, operating currently as Expresso), Scancom Ghana Limited (operating currently as MTN), and One Touch (currently operating as Vodafone). With the recent licensing of Airtel and Glo, the total number of operators currently stands at six – one of the highest on the continent. While initially slow, the mobile telephony sector in Ghana has witnessed remarkable growth, rising from over 100,000 subscribers to the various networks in 2000 to over 25 million by 2012, according to International Telecommunication Union data sources.

There has also been a corresponding rise in the number of young people engaged in informal support businesses in the sector. Even though these businesses are located across Accra, five localities were selected for this study (see Figure 11.1). Kwame Nkrumah Circle and Madina were selected as the mapping exercise, conducted as part of the data collection, revealed these localities as being hubs or clusters for mobile telephony businesses in Accra. The other three localities were selected due to having a concentration of specific

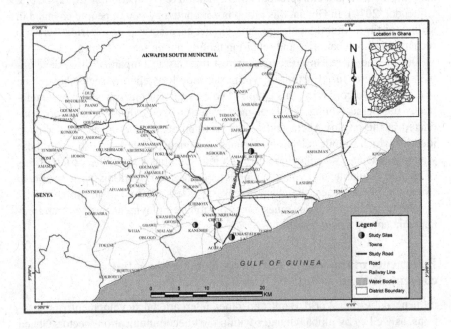

Figure 11.1 Map showing study sites in the Greater Accra Metropolitan Area (source: RSGIS Lab, University of Ghana).

mobile telephony business activities: at Kaneshie lorry station there are many businesses engaged in mobile phone battery recharging; Tema lorry station is a good example of a transport terminal that accommodates the mobile phone accessories trade conducted on push-trucks; and the Legon–Madina road represents a typical busy road network where young people engage in mobile telephony airtime street trading.

Methodology

The chapter is based on a broader study titled *Youth Livelihoods and Entrepreneurship in the Mobile Telephony Sector in the Greater Accra Metropolitan Area* (Afutu-Kotey, 2013). Using various field instruments – questionnaires, life-trajectory interviews, biographical interviews, and in-depth interviews – primary data was collected from the five study localities above and from institutional sources. The questionnaire survey started by mapping and listing all business activities in four of the study localities,[2] followed by stratified sampling and the use of a random number generator in Excel. In all, 354 young people aged 15–35 were interviewed. Issues covered by the survey included, among others, their current business activities, motivations for starting these businesses, financing, skills acquisition, and support for business operations.

Life-trajectory interviews were conducted with 15 young people purposively selected in order to obtain a viable representation of the various business types in the sector, the gender dynamics involved, and the study locality. These individuals were interviewed once every four months over a period of three years. The aim of the interviews was to monitor changing trends with respect to business activities, support, and growth. In addition, 25 biographical interviews were conducted, covering issues similar to the life-trajectory interviews.

In order to obtain information on legislation, policies, and programmes regarding the mobile telephony sector and the informal economy generally, 13 key informants were interviewed. These included officials from two commercial banks, two microfinance institutions, the Accra Metropolitan Authority, and the Madina Zonal Council. Others included four knowledgeable entrepreneurs with at least 10 years' working experience in the mobile telephony sector and two executives of the Tip-Toe Lane Traders Association, located in the Kwame Nkrumah Circle in Accra.

Unravelling youth engagement in different mobile telephony businesses

Over two-thirds of the entrepreneurs interviewed were young men, highlighting how the mobile telephony sector is male-dominated. This ties in with recent studies that indicate the gendered nature of work in the informal economy in Ghana (Ghana Trades Union Congress, 2011; Gough, Tipple, & Napier, 2003; Langevang & Gough, 2009) where an increasing number of men work, although women still dominate the sector overall (Overå, 2008; Wrigley-Asante, 2008).

Although Ghanaians clearly constitute the majority of the young people working in the mobile telephony sector (87%), there is a Nigerian presence in the sector (13%) in line with the increasing flow of Nigerians into different sectors of the Ghanaian economy in recent times.

Generally, education levels among young people in the sector are relatively high and this transcends gender barriers. In total, 45% have completed senior high school (SHS), while 33% have completed junior high school (JHS), and only 6% have no educational qualifications. Although about 3% of respondents have completed tertiary education or its equivalent, this group comprises only young men. The observed trend in educational qualifications falls in line with the general trend within the Ghanaian population of increasing numbers of young people who have successfully completed either primary school, JHS or SHS, but cannot find further formal educational or training opportunities to foster their educational aspirations (Palmer, 2009).

The study identifies five main types of businesses that engage young people in the mobile telephony sector: trading in mobile phones, airtime or top-up units, mobile phone repair, recharging of phone batteries, and mobile phone accessories. The following section describes young people's engagement in each of these business types.

Mobile phone sales

In the early years of the sector's emergence in Ghana, mobile phone use was limited mainly to the few elite groups that could afford the new technology – politicians, business people, and others who wanted to position themselves in a particular social class. As competition in the sector intensified with the licensing of additional operators, the cost of operating a mobile phone decreased significantly as service operators shifted their emphasis from contract or post-paid lines to pre-paid or pay-as-you-go lines, thus widening the customer base. With increasing taste for the new technology among Ghanaians, young business people identified an opportunity in the sector. Many Ghanaians travelled to Europe and Asia to import mobile phones. Some of these importers included young university students who would occasionally transform themselves into seasonal entrepreneurs on return from vacation in Europe and import mobile phones into Ghana. As the demand for the new technology increased, these young traders brought in new and used mobile phones.

Currently, the sale of mobile phones, both new and used, is an important economic activity for many young people in Ghana. Spatially, although the mobile phone trade is widespread in many parts of the country, the Kwame Nkrumah Circle area has become the hub of the mobile phone trade in Accra. Mobile phone sales constitute over 45% of the sampled population. The business is heavily dominated by men, who account for 91% of the mobile phone trade in all the study localities. The limited number of women in the phone sales business can be attributed to the general perception among young people that business in the mobile telephony sector is a "male" activity. This phenomenon is captured by 28 year-old Stella, who trades in airtime:

The phone business is for men. I don't have the money to invest in phones or accessories. Then again, some of the phones may be faulty and I don't have the expertise in repairing phones. Whenever you see a shop that deals in either phones or accessories, there may be a male repairer alongside.

Stella's statement shows the extent to which both financing and skills requirements and expertise structure the gendered nature of businesses in the mobile telephony sector. This notwithstanding, phone businesses range from very small enterprises operated by hand or from trays to tabletop units and glass cabinets, car booths, and shops, which represent the higher echelons of the phone business (see Table 11.1).

At the lower end, women engage in selling mobile phones by displaying their goods on trays while their male counterparts sell by hand. Many young people enter the business by selling phones by hand or from trays or tabletops and then gradually expanding their enterprises into shops and containers. Some report having started their business on perceiving and exploiting opportunities in the phone trade. The majority of these people were already involved in other trading activities, which gave them a chance to accumulate capital prior to starting a phone business. This scenario is exemplified by the case of Eric (aged 33) who started by importing mobile phone accessories and later diversified into mobile phone imports. He explained his reasons for establishing a business in the mobile telephony sector as follows:

I was selling shoes before going into the phone business. When the mobile phone companies started introducing phones into the market, I realised the phones they were bringing were much bigger, and people wanted smaller phones, so I started importing smaller phones which very much caught on among people.

Like many young people, Eric was self-employed, dealing in shoes before perceiving an opportunity in the mobile phone sector. He paid particular attention to trends in the sector, identified opportunities in areas where he thought his business would thrive, and exploited opportunities by going into mobile phone imports. Eric was able to utilise his savings and capital from his previous business to establish his phone business.

Some young people are able to establish phone businesses using resources embedded within their social relations, as exemplified by the case of Albert (aged 27). Prior to establishing his business, Albert worked with his brother for almost two years. This gave Albert the needed expertise and knowledge of the sector as well as an opportunity to accumulate some savings to establish his business. He noted that, "My elder sister who is abroad sent me US$500 which I used in addition to my savings to establish this business." Albert's case, like that of many of the young people studied, demonstrates the significance of social capital with respect to access to resources for business among young people. This observation echoes other studies on Southeast Asia and Africa, which

Table 11.1 Mobile telephony business types and operational size

Activity	Size of operation (structure type)						
	Permanent		Temporary				None
	Shop	Table top	Kiosk	Glass cabinet or box	Push-truck	Car booth	By hand
Sale of mobile phones	X	X		X		X	X
Airtime	X	X	X	X			X
Mobile phone repair	X	X	X				
Mobile phone battery recharging	X	X					
Sale of mobile phone accessories	X	X	X	X	X		

Source: Mapping exercise, 2010/2011.

identify the influential role of social capital especially among young people as they go about the process of establishing and running a business (Turner, 2007; Turner & Nguyen, 2005; see also Chapter 15).

The perceived lucrativeness of the phone business as against other business types in the mobile telephony sector, such as airtime, is the main motivation for a section of the young people surveyed. It is, therefore, common to see young people who started by trading in the airtime business, for instance, working their way up the ladder by saving and expanding their operations into perceived high-value activities such as the mobile phone and accessories trade. The case of 23-year-old Patricia illustrates this. Patricia started by selling airtime "in order to earn more money", but after two years, not satisfied with the income she was earning, she diversified into phone sales because she perceived it would offer opportunities to earn a higher income.

Mobile phone airtime or top-up units

During the initial stages of the mobile phone revolution in Ghana, post-paid customers outnumbered pre-paid ones; over time, the latter have overtaken the former. Hahn and Kibora (2008) identify the limited economic capacity of most mobile phone owners as the main reason for the high popularity of pre-paid customers. Added to this is the informal nature of economic transactions in Ghana and the difficulty in identifying and locating people, which contributes to the popularity of pre-paid customers.[3] Thus in order to make calls most mobile phone users have to buy airtime or top-up units, which has generated business opportunities for young people who sell the scratch cards that contain the units. The airtime trade has a heavy female presence with about 52% of all women in the sample engaged in airtime trade compared to just 10% of all men. This can be attributed to the low capital requirements for start-up and the fact that no specialised skills are required. With about GH₵50, it is possible to start a mobile phone airtime business.[4] Although some wholesale activities were observed, most young women were engaged at the retail end of the airtime value chain selling mainly from tabletops, glass cabinets, or boxes or by hand along the main roads, at major traffic intersections, lorry stations, markets, schools, and in neighbourhoods.

In a few instances, young people did not need start-up capital to initiate an enterprise. Patience (aged 31), for example, started her business through MTN under a scheme where airtime vendors register with the company and are supplied airtime at wholesale prices, which they sell and repay post-sales.[5] To complete the registration process, potential vendors have to offer a guarantor. In Patricia's case, a good friend who had first told her about the scheme acted as her guarantor. Patricia noted the significance of this scheme as follows: "I never had money to trade. Since I completed the forms with MTN, I now have my own business." Although her case demonstrates the existence of a working relationship between network operators in the formal sector and young people working in the informal sector, generally there is limited support from formal organisations for young people operating informal businesses in the mobile telephony sector.

Structural constraints, typified by the generally difficult economic situation in Ghana which offers limited employment prospects for young people, appears to be the dominant reason for going into airtime businesses among the majority of youth. The case of Kofi (aged 22), who had completed JHS and currently trades in airtime, exemplifies how these constraints combine to influence young people's decision to establish such a business. He explained why he went into the airtime trade as follows: "There are no jobs for us [young people] to do so when you get something doing that can take care of you, you have to hold on to it well." For a section of young people, the need to save money for further education is their main reason for going into the airtime business. A typical instance is Sedro (aged 22) who stated: "I am doing this business because I want to save some amount of money in order to go back to school." Running an airtime business for such people is only a stopgap arrangement that enables them to save to finance their education. Other young people sell airtime alongside studying. Sarah (aged 22) explained how "Currently, I am at Ideal College writing my Science and Maths. As a student I needed money to take care of myself and this is the reason why I started this business." Thus, running an airtime enterprise offers avenues for young people to save in order to further their educational aspirations and sustain themselves while at school.

Mobile phone repair

For many Ghanaians, acquiring a mobile phone is an expensive investment. Having bought a handset, however, people are willing to go the extra mile to maintain it. The need to fix or replace broken parts has thus generated business avenues for young people in the phone repair sector, which is male-dominated with the entire sample constituting young men. Some of the repairers surveyed had formerly engaged in other trades such as the repair of watches, radios, and other electronic devices. The watch repair business is no longer as lucrative given that people now use their phones to read the time, and mobile phone repair is considered to be more profitable than repairing other electronic goods. The young people who move from these businesses into mobile phone repairs tend to be the most skilled as their prior technical knowledge makes it easier to acquire the necessary skills. Some young people did, however, pass through an informal apprenticeship system under the tutelage of experienced repairers in order to acquire the requisite skills. These apprenticeships usually lasted between two and six months for individuals with some background in electronics and about a year for those without.

Aside from acquiring skills informally, some young people take up training opportunities offered by the company Rlg Communications in mobile phone repair.[6] However, among the 10 young people who were tracked after they had successfully undergone the training programme, only two succeeded in establishing a mobile phone repair business. Most of the young people noted that, prior to the training programme, they had been promised support by the recruiting nongovernment organisation (NGO), but this had yet to materialise. Bernard (aged 17), who went through 6 months of Rlg training, explained that, "While training, they [the NGO] told us they would give us all the tools and equipment and also help us establish our own businesses, but unfortunately, since we completed the training, they have not given us the tools." With the limited institutional – and family – support available, Bernard has not been able to mobilise enough capital to establish a phone repair business. His case parallels that of many young people who go through various training programmes with high expectations, but find it difficult to reconcile these expectations with jobs or employment outcomes post-training (Darkwah, 2013).

The kinds of services offered by mobile phone repairers vary from the complex "decoding" or "unlocking" phones to the simple technological replacement or repair of broken parts, such as screens and microphone units. Some imported phones, especially from Europe and North America, are coded or locked to specific networks in order to restrict access only to these networks (Hahn & Kibora, 2008). The young people in the decoding business use microcomputers with specialised software that enables them to decode these phones.

A strong sense of passion and interest intertwined with previously acquired skills and identified opportunities underpin most young people's decision to go into business in the mobile phone repair sector. A typical example is Kojo (aged 28), who apprenticed in computer hardware technology after completing SHS. He explained that, although he occasionally works as a computer technician, his main line of business at present is mobile phone repair. He established the enterprise because of his strong passion for mobile phones: "I am very much interested in mobile phones and because of my skills and knowledge in computers I got interested in repairing mobile phones." In many instances, this enthusiasm is influenced by the fact that young people see mobile phones as a technological device for their generation. Samuel (aged 24) noted that "The mobile phone is for our generation and it has come to stay. We have to make the most of the technology."

Mobile phone battery recharging

Some young people have identified a niche in the mobile telephony sector for mobile phone battery recharging businesses. These tend to be operated by young men from tabletops and be located near major transport terminals and markets (see Figure 11.2). The target group includes traders and their customers as well as drivers, their assistants, and other passengers who tend not to have access to electricity for most of the day to recharge their phones. Many Ghanaians use cheap phones that have a short battery life and need to be recharged often. Given all these factors, young entrepreneurs have established their mobile phone battery recharging businesses in busy places to meet the demand.

The operation involves using conventional plug-in cords and single docking stations connected to a power base. The phone batteries are then lodged in the

Figure 11.2 Phone recharging business at Kaneshie lorry station, Accra (photo: Robert
    Afutu-Kotey).

docking station until fully recharged. Often, the operation comprises a small
table with only a few plug-in cords and a single docking station, but can grow to
a set of bigger tables accommodating between 35 and 50 recharging equipment
sets. Operationally, recharging a phone or battery costs GH₵0.50, and, on
average, young people are able to recharge between 45 and 60 phones a day.
They pay between GH₵1 and GH₵2 daily to nearby landlords for the cost of the
electric power supply. To prevent fraudsters from laying claim to phones, the
business operators use a manual coding system to identify each customer's
phone which is shared only with the phone owner. To collect their recharged
phones, owners must identify their instrument by the code issued to them. The
greatest risk to the business is if a third party obtains access to the code and uses
it to collect a phone fraudulently. When this happens, the business operator has
to replace the owner's phone (which may end up with the police), thereby losing
money.

For the majority (60%) of young people, their main reason for going into the
phone recharging business is the limited availability of other employment opportunities. For others, however, there is the prospect of earning a higher income than
from other businesses outside the mobile telephony sector. This is exemplified by
the case of 23-year-old Ofori who has been in the phone recharging business for

close to two years. He explained how "A friend introduced me into the business. He told me that the business is profitable. I was selling handkerchiefs and I used the capital to establish this business. It has been good so far." Although his desire to earn more explains his decision to enter the phone recharging business, Ofori's statement also demonstrates the significance of social capital among young people.

Mobile phone accessories

As people's mobile phones age, the desire to maintain their appearance and functionality rises, given that possessing a mobile phone is considered something of a status symbol among a section of society. Some people also find it difficult to replace either old or faulty phones because they cannot afford new ones. These activities have generated business opportunities for young people in the phone accessories trade, which includes the sale of phone accessories and replacement parts such as batteries, chargers, ear pieces, covers, and screen protectors. These businesses are concentrated in many parts of Accra, but particularly in Madina, Kwame Nkrumah Circle, and Kaneshie and Tema lorry stations. Phone accessories businesses often operate from tabletops, glass cabinets, and push-trucks or from shops and freight containers.

The youth engaged in this business at the Tema lorry station in central Accra do not have fixed posts from where they ply their trade. Instead, they have acquired push-trucks on which they display their products at the terminal. Constrained by the lack of business operating space, these young people occupy temporarily vacant parking lots at the station and move their push-trucks with phone accessories in and out of the lots as needed (see Figure 11.3). The main customers are passengers at the lorry station and other individuals who have come there to trade. Interestingly, these young people have become an integral part of the station and pay a fee to the station authorities towards its maintenance and upkeep.

For most young people, their main reason for going into the accessories business is the desire to earn relatively more income. This is demonstrated by the case of Opoku (aged 26), who has been trading in phone accessories at the Tema lorry station for the last three years. He noted that, "I have a friend who is into phone accessories and he advised me to go into the business because it is good business." Opoku used to trade in electronic appliances such as radios, watches, and alarm clocks, but does not regret having invested in the phone accessories business because "phone accessories move faster than the electronic products I was dealing in". Similarly, Sammy (aged 31) was trading in handkerchiefs and perfumes when he realised that the proliferation of fake perfumes had made the trade less lucrative. He explained, "I used my capital to establish this business after being introduced into it by a friend." With the mobile phone revolution, many young people then trading in other products perceived opportunities in the mobile phone accessories business and used their operating capital generated elsewhere to exploit this new potential.

Figure 11.3 Phone accessories business at Tema lorry station, Accra (photo: Robert Afutu-Kotey).

Conclusion

The growing significance of the mobile phone in African cities has attracted the attention of the research community, although little attention has been devoted to informal entrepreneurial businesses in the mobile telephony sector (Chiumbu & Nyamanhindi, 2012). This chapter has uncovered the role of informal businesses as avenues for generating youth employment in Accra. Young people have identified and are exploiting opportunities in the mobile phone and airtime trades, in mobile phone repair, phone battery recharging, and mobile phone accessories. The sector is male-dominated, though young women dominate in the sale of airtime, which is the easiest subsector to enter.

Young people's motivation for going into business in the mobile telephony sector tends to be distinct with respect to business type. A key driver of the decision to enter the phone and accessories businesses is their perceived profitability. A sense of passion and interest, intertwined with previously acquired skills, seems to be the main reason attracting other young people to the phone repair sector. In relation to airtime trade, the desire to finance their schooling or build a capital base for future business plans drives many young people to engage in this activity, which is often only a stopgap business. Irrespective of the precise nature of the business within the sector, many of the young entrepreneurs are not static business operators but move up the hierarchy from perceived low-return businesses such as airtime to relatively high-return ones such as phone and accessories sales. Generally, young people appear to want to remain engaged in the mobile telephony sector and most respondents expressed positive aspirations with respect to their business, ranging from learning new skills and acquiring new equipment with which to run their business, to engaging in the import and wholesale of mobile phones and related products. Only a few expressed a desire to quit the sector – primarily those involved in the airtime trade who intend to mobilise their savings to return to full-time education.

Overall, social capital appears to be critical for many young people as a source of support for establishing and running a business. However, most young people have weak access to institutional support as they go about the process of identifying and exploiting informal business opportunities in the mobile telephony sector (see also Ghana Trades Union Congress, 2011). In an attempt to address these shortcomings, government policy should recognise the contribution of young people running various informal economy activities through their engagement in production and distribution mechanisms as well as through the taxes and other fees they pay. Also, training programmes based on the real needs of young people, for instance business management training within an informal economy context, could be taken up by governmental agencies (such as the National Board for Small Scale Industries), NGOs, and the private sector. This would go a long way in propelling growth and sustainability of the businesses owned by the youth.

Notes

1 Donner (2008), for instance, has reviewed roughly 200 studies in the developing world, most of which generally focus on the determinants of mobile phone adoption, the impact of mobile phones at the macro- and micro-levels, and their symbolic nature.
2 The listing exercise did not include the Legon–Madina Road where young people run ambulatory operations.
3 The unbanked population in Ghana is over 80%, according to PricewaterhouseCoopers (2011).
4 At the time of collecting the data, the exchange rate was GH₵0.6 to US$1.
5 MTN is the biggest mobile telephony operator in Ghana with a subscriber base of about 11 million.
6 Rlg Communications is a Ghanaian-owned limited liability company engaged in the production of communications equipment such as mobile handsets, electronic notebooks, tablets, laptops, and LCD TV monitors, among others.

References

Afutu-Kotey, R. L. (2013). *Youth livelihoods and entrepreneurship in the mobile telephony sector in the Greater Accra Metropolitan Area.* Unpublished doctoral thesis, University of Ghana, Legon, Ghana.
Aker, J., & Mbiti, I. (2010). *Mobile phones and economic development in Africa* (Working Paper No. 211). Washington, DC: Centre for Global Development.

Chiumbu, S., & Nyamanhindi, R. (2012). Negotiating the crisis: Mobile phones and the informal economy in Zimbabwe. In S. Chiumbu & M. Musemwa (Eds), *Crisis! What crisis? The multiple dimensions of the Zimbabwean crisis* (chap. 3). Cape Town: HSRC Press.

Darkwah, A. K. (2013). Keeping hope alive: An analysis of training opportunities for Ghanaian youth in the emerging oil and gas industry. *International Development Planning Review, 35*(2), 119–134.

Donner, J. (2008). Shrinking fourth world? Mobiles, development, and inclusion. In J. E. Katz (Ed.), *Handbook of mobile communication studies* (pp. 29–42). Cambridge, MA: MIT Press.

Etzo, S., & Collender, G. (2010). The mobile phone "revolution" in Africa: Rhetoric or reality? *African Affairs, 109*(437), 659–668.

Frempong, G., Esselar, S., & Stork, C. (2005). Ghana. In A. Gillwald (Ed.), *Towards an African e-index: Household and individual ICT access and usage across 10 African countries* (chap. 6). Johannesburg: Wits University School of Public and Development Management.

Ghana Trades Union Congress. (2011). *The state of the urban informal economy: A focus on street vendors*. Accra: Ghana Trades Union Congress.

Gough, K. V., Tipple, A. G., & Napier, M. (2003). Making a living in African cities: The role of home-based enterprises in Accra and Pretoria. *International Planning Studies, 8*(4), 253–277.

Hahn, H. P., & Kibora, L. (2008). The domestication of the mobile phone: Oral society and new ICT in Burkina Faso. *Journal of Modern African Studies, 46*(1), 87–109.

Langevang, T., & Gough, K. V. (2009). Surviving through movement: The mobility of urban youth in Ghana. *Social and Cultural Geography, 10*(7), 741–756.

Langevang, T., & Gough, K. V. (2012). Diverging pathways: Young female employment and entrepreneurship in sub-Saharan Africa. *The Geographical Journal, 178*(3), 242–252.

Overå, R. (2008). Mobile traders and mobile phones in Ghana. In J. E. Katz (Ed.), *Handbook of mobile communication studies* (pp. 43–54). Cambridge, MA: MIT Press.

Palmer, R. (2009). Formalising the informal: Ghana's National Apprenticeship Programme. *Journal of Vocational Education and Training, 61*(1), 67–83.

PricewaterhouseCoopers. (2011). *The Ghana banking survey 2011: Sustaining growth – Challenges and opportunities*. Accra: PwC.

Turner, S. (2007). Small-scale enterprise livelihoods and social capital in eastern Indonesia: Ethnic embeddedness and exclusion. *The Professional Geographer, 59*(4), 407–420.

Turner, S., & Nguyen, P. A. (2005). Young entrepreneurs, social capital and Doi Moi in Hanoi, Vietnam. *Urban Studies, 42*(10), 1693–1710.

Wallsten, S. J. (1999). *An empirical analysis of competition, privatization, and regulation in Africa and Latin America*. Unpublished manuscript, World Bank, Washington, DC. Retrieved from http://elibrary.worldbank.org/doi/pdf/10.1596/1813-9450-2136.

Wrigley-Asante, C. (2008). Men are poor but women are poorer: Gendered poverty and survival strategies in the Dangme West District of Ghana. *Norsk Geografisk Tidsskrift – Norwegian Journal of Geography, 62*, 139–148.

12 Youth entrepreneurship in a small-scale gold mining settlement in Ghana

Marshall Kala

Introduction

Small-scale mining (SSM) has long existed in Ghana but has become more widespread in recent years linked to rising gold prices and the declining interest in agricultural activities, especially among the youth. Young people from a wide range of backgrounds and both genders now see SSM and associated support activities as viable livelihood options. The sector remains highly controversial, however, and most small-scale miners operate illegally as the process of obtaining a mining lease is lengthy and tedious (Hilson, 2001). Small-scale miners are locally referred to as *galamsey*, which literally means "gather them and sell". The media has launched a strong campaign against *galamsey*, denouncing them as "criminals, vandals, environmental polluters and self-harmers" (Bush, 2009: 57). The SSM sector has also been widely condemned in policy circles and a government task force has been established to try to curtail *galamsey* activities.

Although there is considerable research on various aspects of mining in Ghana (see, among others, Akabzaa, 2009; Banchirigah, 2008; Bloch & Owusu, 2012; Hilson, Amankwah, & Ofori-Sarpong, 2013; Hilson & Potter, 2005; Larsen, Yankson, & Fold, 2008), little is known about who engages in the different activities that constitute SSM, nor have any studies investigated in detail the ancillary activities that take place in mining settlements. This chapter explores the range of entrepreneurial activities undertaken by youth in SSM. It examines which young people engage in various activities, their motivation for starting a venture, the challenges they face, and their aspirations. The chapter draws on extensive fieldwork conducted in the recently established mining settlement of Kui in northern Ghana. It shows how SSM is an important source of livelihood for many young people, some of whom progress through the ranks within the sector while others use mining as a stepping stone to other income-generating activities.

Gold mining in Ghana

Ghana is endowed with vast mineral deposits that are mined by large-scale mining corporations as well as by SSM units. The latter are defined as operations based on less than 25 acres of land (Hilson, 2001). The liberalisation of the mining sector and the introduction of new technologies in large-scale mining companies has led to labour rationalisation and reduced employment (Aryeetey et al., 2004). Many laid-off miners have moved into SSM, which accounts for about 6% of gold production in Ghana and is estimated to engage 300,000 people. If dependants are included, the sector provides livelihoods to another one million people (Hilson & Banchirigah, 2009; Hilson & Potter, 2005). SSM has spread from the south-west of Ghana, where it has primarily taken place, to the north of the country in recent decades (Gough & Yankson, 2012). Although northern Ghana is primarily agrarian, a combination of structural adjustment policies linked to globalisation, weather-dependent agriculture, and failure to modernise the sector have led to a situation where food-crop farmers are the poorest in the country (Ghana Statistical Service, 2007). Coupled with the generally declining number of job opportunities for youth and colonial neglect, these factors have pushed people to search elsewhere for work and income. Income diversification is thus a widely practised coping strategy among farm households in northern Ghana (Marchetta, 2008).

Kui is a newly established mining settlement located in Bole District in the Northern Region, which did not exist prior to the discovery of gold. The township is located about 20 km off the Kumasi–Wa road along a very rough dirt track that is accessible only by motorbike, truck, or four-wheel-drive vehicle. Gold extraction began in Kui in the early 2000s with rudimentary tools and technology, although chanfan machines, which are used to grind ore, were subsequently introduced. Most operations involve surface mining although deep-pit mining is carried out on a limited scale in the dry season. The chief of Bole, who owns the land, has tasked a chief in Kui to collect taxes from the businesses there. Kui's population is estimated to be 7,000 (Hilson et al., 2013) with people coming from all regions of Ghana as well as Burkina Faso, Nigeria, and other countries. The houses are makeshift structures built primarily of polythene with mud floors; only a few structures have wooden walls or cement floors.

Methodology

Both quantitative and qualitative primary data were collected for this study. The purpose of the quantitative aspect was to obtain an overview of which youth are involved in different activities. The quantitative data was collected using a questionnaire survey administered to 200 youth from the mining settlement, who were randomly selected using a residence-based list of young entrepreneurs compiled by the author. Young entrepreneurs were identified on the basis of screening questions such as "Are you working for yourself or are you being paid by somebody to work for him/her?" and "How old are you?" A person aged 15–35 working for himself/herself or paying people to work for him/her was considered an entrepreneur. Four peer interviewers, who were university graduates, administered the questionnaires alongside the author through face-to-face interviews in whichever language respondents were most comfortable with (their mother tongue, English, or any other language). Each interview lasted approximately 30 minutes and the entire data-collection process took six weeks.

Qualitative methods were also employed to gain a finer understanding of young people's SSM experiences. Five focus group discussions (FGDs) were conducted in Kui to assess why youth engage in SSM and the nature of any institutional support provided. Each FGD had 5–10 participants engaged in a range of activities and lasted approximately one hour. Ten in-depth interviews were also conducted in the settlement, lasting about half an hour each. The objective was to examine young people's work history and the factors driving their decision to migrate to the mining settlement, and to learn more about their businesses and experiences in Kui. All the in-depth interviews and FGDs were audio-recorded and transcribed verbatim. Living in Kui for several weeks also allowed the author to engage in participant observation, which provided insights into the day-to-day challenges that residents face.

Entrepreneurial activities in Kui

As stated by a respondent: "If God gives you the gift of life and you come here, there are several things that you can do." This statement underscores the range of entrepreneurial activities in the mining settlements, which can be divided into two groups: those that involve mining and activities that support mining. People engaged directly in mining include, but are not limited to, chisellers, drillers, pit owners, blast men, haulers, processors, buyers, and sponsors. Activities that support mining include catering services, the sale of provisions, transportation, health services, entertainment, and religious activities. These are analysed in two sections, examining first entrepreneurial activities in mining and then entrepreneurial activities that support mining.

Entrepreneurial activities in mining

SSM starts with the identification of a site believed to contain gold. A pit owner engages diggers – including chisellers, drillers, and blast men – who dig until the pit hits a gold-bearing rock. The rock is removed from the pit by haulers or "locoboys" and transported by head porters. Each pit has a hut, referred to as a ghetto, to shield miners from the elements. Sponsors finance the equipment and explosives and, in some cases, provide food for their workers, although the cost of this is deducted on the sale of the gold. The extracted ore is processed or sold to dealers. The profit from the sale is divided three ways among the underground workers, the pit owner, and the sponsor. These activities are discussed in turn, starting with the extraction process.

The extraction process is the main entry point for young people, especially men. Just over one-quarter (27%) of the young people surveyed were engaged in chiselling, drilling, or blasting (Table 12.1) (see Figure 12.1). Although these activities require a degree of skill, training is provided on the job by more experienced miners. Young people can, therefore, enter the sector without any

Figure 12.1 Young miner in a pit chiselling ore in Kui (photo: Marshall Kala).

prior mining knowledge. Almost all the youth engaged in chiselling and drilling in Kui were male, although there were some young women involved in digging in Kui as most of the mining is surface or "top-face" mining. As expressed in an FGD in Kui, some mining activities are gendered while others are not:

> There are certain jobs that can only be done by men because of the energy required for the job. Ghetto [underground] work is usually done by men.... Machine work can be done by both men and women. In fact, there is no hard and fast rule about who works where.
>
> (Mohammed, aged 29)

Hauling, in which 7% of the miners are engaged (Table 12.1), does not require any special skills. Its capital requirements are very low and it serves as another entry point into the mining profession for many young men. Haulers typically have limited education; what is required is strength and care to help prevent accidents. Similarly, another activity is head porterage, which is used when the load is not too far from the processing site, if the ore is inaccessible to vehicles, and/or if it is cheaper (see Figure 12.2). The young people engaged in head loading make up about 9% of those engaged in mining. Most are women (85%), unlike the case of diggers and locoboys. Some young entrepreneurs negotiate a good price for transporting the ore and then engage others to carry it for a fee. Agnes (aged 30), for example, works as a head porter in Kui, her young son strapped to her back. She lived in a village near Wa prior to her move to Kui, brewing local beer (*pito*) and farming bambara beans to earn a livelihood. Agnes decided to move to Kui after her son was born in order to earn more money after a friend paid her transport fare.

Table 12.1 Entrepreneurial mining activities in Kui

Activity	Percentage
Chiselling/drilling/blasting	27
Hauling/locoboy	7
Head porterage	9
Pit/ghetto owner	7
Scavenging for waste ore	2
Processing ore	38
Dealing	11
Sample size	*115*

Source: Questionnaire survey (2013).

Figure 12.2 Young women carrying ore to processing sites with containers called "bashers" in Kui (photo: Marshall Kala).

The young people engaged in digging, hauling, and porterage aspire to make enough money to transition into other activities that are less taxing and risky, and more lucrative. This is understandable as most of the deaths that occur due to cave-ins, suffocation, and flooding in SSM are associated with these activities. Typical strategies to accumulate sufficient money to move into other activities include saving to invest in a processing machine or to start another business in the mining settlement, in the closest large settlement, or in their home town. Zakaria (aged 34), for example, moved to Kui after her divorce. She is a mother of three who wanted to move into the services economy but did not have the means to do so. Zakaria carried ore for three months to raise GH₵150 (US$50), which she then used to start a catering business in Kui. She serves *tuozaafi*, a local maize meal popular among people in northern Ghana, bambara beans, and rice. Zakaria's story was corroborated by Area (aged 32) in an FGD:

> If you don't have money, you have to toil [in mining] until you get a few
> cedis. You can then invest that in a few drums of diesel or petrol for retail.
> If you are fortunate, you may eventually start a store with the accumulated
> profit.

These cases illustrate how, for some young people, working in the mining sector
is a short-term activity and part of a deliberate strategy to generate capital to
establish businesses outside the sector.

Pit owners are individuals or groups of individuals who make the initial
investment in a pit in the hope of hitting gold-bearing rock. Given the capital
and experience required, the few pit owners are all male. It is rare for young men
to enter mining as pit owners, rather it is a role they aspire to achieve. Justice
(aged 34) traced his path to becoming a pit owner in Kenyasi:

> I moved from Accra to Tarkwa before moving to Kenyasi. In Tarkwa, I
> worked with Bogoso Gold Limited. I worked on contract as a welder for
> three years and left. When I came here in 2007, I used to go down to chisel
> and drill but now I have many workers and supervise them. I was told about
> this place by a friend when I was in Tarkwa. I changed from chiselling and
> drilling when I made money and hired more hands. I used the money I made
> from Tarkwa as well as support from friends.

This shows how some miners save money from working in large-scale mines
before moving into SSM in the hope of making more money and then moving
through the ranks within SSM. Justice has spent more than GH₵300,000 in the
past two years just to get to gold-bearing rock. The key challenge pit owners
face is the lack of geological and other technical information, hence they take
major risks when deciding on where to sink a pit. In Kui, some pit owners complained of having processed a Kia-truckload of ore and not having made enough
to even pay for the services of the truck. Some youth even consult spiritualists or
mallams who sacrifice white doves as a way of trying to increase their luck.

A small number of youth scavenge waste ore from the mining to process for
gold. They are considered to be at the bottom of the mining hierarchy and comprise mostly uneducated women from the north of Ghana. Young men were not
interested in scavenging or thought they could earn more money elsewhere. The
young women, however, do manage to eke out a living from this activity.

Ore processing is another essential component of mining. It consists of a
range of activities, including crushing and grinding the ore into powder, adding
water to form slurry, washing it with jute materials or carpets to sieve out the
gold, and then using mercury to amalgamate the gold particles. Mining in Kui is
alluvial and so the ore is in the form of gravel. A chanfan machine is used to
grind the ore to liberate the gold particles. In all, 17% of respondents were
engaged in processing the ore, an overwhelming majority of whom (95.7%)
were male. Women used to pound the ore into powder by hand but this is now
done by machine, operating which is seen as being the sole preserve of men;

women's role in processing has thus diminished greatly. Only a couple of women owned processing machines and had to engage men to operate them. The owner of the ore usually pays half the processing cost, the rest is paid on completion of the work. Processing does not require considerable experience or training; what is needed is the necessary finance to purchase a processing machine. As the machines cost a considerable sum, typically GH₵3,000, there are instances of people who had saved money working in Europe and returned to Ghana to engage in processing. Others had managed to save from digging or other businesses in Ghana.

A number of hazards are associated with processing, which were highlighted by the young male chanfan operators during an FGD. They reported often suffering hearing loss because the machines were very loud and they lacked ear protection equipment. Chanfan belts can also cause injuries such as severed fingers and lacerations, especially when operators tire from working around the clock. The "smooth machines", on the other hand, generate a great deal of dust, which affects breathing. Many of these hazards could be avoided if those engaged in processing were to use personal protection equipment. However, this is rarely the case in SSM and this compounds the hazards more frequently mentioned in the literature, such as the use of mercury during the amalgamation process (Aryee, Ntibery, & Atorkui, 2003; Nyame & Blocher, 2010).

Dealers, who made up 11% of the sample, can be subdivided into sponsors and buyers, though as many individuals assume both roles they were grouped together for the purposes of this study. Sponsors – all of whom were male in this case, barring one – play an important role in the mining process because they are its principle financiers. In Kui, sponsors enter the process when an area is assayed and judged to possess rich deposits. They are heavily relied on by operators when the need for money arises, be it work-related or for personal reasons. Becoming a sponsor requires both capital and contacts.

Buyers tend to be located at the mining site and usually buy the gold in small quantities from the miners for onward sale to other buyers in bigger towns such as Tinga, Techiman, and Bole. In Kui, where the sponsors are also buyers, miners have to sell the gold to their sponsor at a rate below the market price. Like sponsors, buyers must also have considerable financial standing and have often generated their start-up capital from previous businesses or employment. This includes many successful miners who have managed to transition from digging, processing, and other mining activities to dealership. The need for operational funding is often obtained through business networks. In Kui, for example, an association of gold buyers from the Gonja ethnic group contribute to a revolving fund every month. The gold dealers are generally more educated than other actors in SSM: most had completed senior high school or tertiary education (52%) or at least junior high school (31%). Bryceson and Jønsson's (2010) study on Tanzania similarly shows that, among other things, having a comparatively high level of education helps people climb up the career pyramid in SSM.

Entrepreneurial activities supporting mining

Although Kui is a newly established mining settlement, there are many activities taking place that are not directly part of the mining process but support miners in all aspects of their daily lives (Table 12.2). The most common support activity is trading, whether from a shop or tabletop or through hawking. Despite the difficulty of transporting goods to Kui, it is possible to buy almost anything there, including provisions, clothing, bags, mobile phones, fuel, and chanfan parts. The entrepreneurial shop owners who have established businesses in Kui usually travel to Techiman to buy Kia-truckloads of items. Although trading usually falls under the purview of women in Ghana (Gough, Tipple, & Napier, 2003; Langevang & Gough, 2009; Overå, 2008), the majority of shop owners in Kui are men (79%). This highlights the atypical nature of new mining settlements and the opportunities that are being seized by some young men, despite the stigma attached to men trading.

Some shop owners act as wholesalers for smaller retailers (petty traders) who sell from tabletops or work as itinerant sellers. Most petty traders are female (71%). Aramata Osman, aged 28, is one such petty trader, who moved to Kui in 2009 with her husband. She spent her first year helping her husband carry ore at his chanfan site. After managing to save some money and with the support of her husband, she started her own business, selling basic foodstuffs, in 2010. Aramata explained how doing business in Kui was much better than elsewhere as it was done on a cash-and-carry basis, i.e. people have to pay up front.

Another support activity that is crucial to miners is the provision of catering services – the livelihood activity of about 16% of respondents, most of whom (78%) were female. Some young male entrepreneurs, however, have seized the opportunity to venture into catering. Kojo Prince (aged 23) makes fast food in Kui. He believes his business is thriving because he is a man who has ventured into a traditionally female domain. Due to its relative isolation, a large company has also drilled a borehole and produces sachet water for the settlement, which individual entrepreneurs then sell alongside other soft drinks. Since entertainment is an integral part of life in the mining settlements, scores of youth operate bars (known locally as drinking spots), the majority (60%) of which are run by women.

Table 12.2 Entrepreneurial activities supporting mining in Kui

Activities that support mining	Percentage
Shop owner/trading	40
Catering service/drinks	16
Repair work (trucks/motors/generators)	16
Other services (photography, carpentry, teaching, etc.)	24
Transportation	4
Sample size	*85*

Source: Questionnaire survey (2013).

Repair work is another crucial service activity that some young entrepreneurs (16% of the sample) have established in Kui. Some of this is directly related to mining activities, such as repairing broken-down machines used to cart and process ore and generators that provide energy for the machines. Others are more standard auto-mechanics who mainly repair the Kia trucks. These mining-related service activities are the sole preserve of men, most of whom had the required skills before starting their business, while a few learned on the job. Other service activities are also widespread in Kui. Some entrepreneurial individuals have drilled boreholes, including one who has established a public bathhouse fitted with showers. Other services offered include photography, teaching at the basic school, carpentry, and charging mobile phone batteries, all of which are dominated by men.

Finally, one of the key services offered by a relatively small number of young people (4%) – all of whom were male – is the transport of both miners and gold. Even though the commercial use of motorbikes or *okada* is illegal in Ghana, it is one of the main means of transport in Kui from the mining site to the main road, a journey of about one hour along a very rough track. However, a directive from the chief forbidding their use has resulted in the number of drivers dropping from over 50 to about 10. Consequently, Kia trucks have become the main form of transportation to and from Kui; in the rainy season, they are the only form of transport as the dire state of the road makes it impassable for smaller vehicles.

Unlike some activities in mining that do not require much start-up capital, most supporting activities require considerable sums to purchase the required inputs and, in some instances, to set up business premises. This start-up capital was obtained mainly from personal savings from previous jobs (52%) or through support from family members (26%) or friends (12%). The dominance of personal savings as a source of start-up capital is underscored by the fact that the youth in this sector had previously worked elsewhere. Overall, young men were more successful than young women in obtaining support from family members or friends, which illustrates their wider networks as well as their greater need for capital, given that their businesses tend to be larger. A challenge that all entrepreneurs face, however, is the fluctuation in business activity linked to the seasonal nature of mining in Kui. Hakeem, one of the leading fuel dealers in Kui, reported that he could earn as much as GH₵30,000 a day in the rainy season when mining activities peak, but might make just GH₵500 during the dry season.

Business owners in Kui face another challenge in that their activities are criminalised because the settlement is not officially recognised. Despite already being heavily taxed by the local chief and his committee, many entrepreneurs said they were prepared to pay government taxes if this would legitimise their activities. Some indicated that they had curtailed their investment in the settlement due to the sense of insecurity, which was heightened when they heard that government officials were clamping down on *galamsey*. The manager of a savings scheme in Kui explained that almost everyone who had deposited their savings with him had withdrawn their money on hearing the news. Armed robbery is yet another challenge, especially when travelling to and from the settlement. Robbers remain undeterred, even after the chief of Kui hired armed guards to patrol the road. These factors highlight how insecurity in mining settlements such as Kui – established purely around mining – can be severe for both miners and entrepreneurs engaged in support activities.

Many entrepreneurs in mining settlements do not limit themselves to one activity but engage in multiple businesses as a way of maximising their incomes and spreading risk. An example is Salifu (aged 32) who owns a provisions shop and diesel station and also sells chanfan parts. He had operated a filling station and shop before moving to Kui, but began dealing in chanfan parts when he discovered they were in high demand there. This way, the demand arising from the congregation of so many people in Kui offsets the slow market in Salifu's hometown resulting from competition from other fuel stations. Other young people combine activities at different times of the year. Alhassan Hakeem is a 21-year-old who dropped out of senior high school because he could not pay the fees and followed his brother, who was processing ore in Kui. He now operates a chanfan machine during the rainy season and teaches at the school in the dry season when there is no water for processing. This way, he is able to work all year round in Kui. Other young people combine mining activities with farming, either close to the mining settlement or back in their hometowns, while some entrepreneurs combine mining and closely linked support activities.

These examples illustrate how mining and support services are not mutually exclusive, and how people combine and switch activities within and beyond the mining settlements as and when necessary. Many of the young people engaged in businesses that support mining plan to continue in this line of business in the long term. A considerable number (44%) plan to start another business outside the mining settlement in the near future, while others are saving to go back to school (6%) or travel abroad (2%).

Conclusion

The importance of the SSM sector in providing youth with livelihoods has been widely recognised by researchers and international organisations such as the World Bank. This chapter has shown how wide-ranging these activities are and how diverse the young people living and working in mining settlements are. They have different backgrounds by way of education, skills, and experience, and participate in different aspects of the mining process based on the kinds of resources they command or can obtain. The activities in this sector are diverse and labour-intensive, and generally offer higher incomes to both young men and women than farming or running a business in other places.

The key motivation for youth entering SSM is the need to better their lives. Many start off by engaging in activities that do not require capital or sell their labour and, over time, start their own business. Obtaining support from formal financial institutions and government institutions in a setting such as Kui is impossible. Most youth depend on their personal savings (earned from entrepreneurial endeavours) and many of those who have set up businesses in the mining settlement or elsewhere financed their start-ups from money realised in mining activities. Engaging in an illegal activity (*galamsey*) and living in an illegal settlement can be risky, as this chapter has highlighted. That young people choose to do so and persevere against the odds is an indication of their entrepreneurial spirit. Rather than being seen in a negative light and condemned within policy circles, this study joins Hilson et al. (2013) in their claim that a supported SSM sector could help reduce unemployment among youth. With the right policy measures, some of the sector's challenges could be addressed so that its full benefits are realised. In view of SSM's potential to absorb youth looking for work, the registration process should be made easier and cheaper to enable more people to legalise their activities, which would then open doors for other forms of support.

References

African Union. (2009). *Africa mining vision*. Addis Ababa: African Union. Retrieved from www.africaminingvision.org/amv_resources/AMV/Africa_Mining_Vision_English.pdf.

Akabzaa, T. (2009). Mining in Ghana: Implications for national economic development and poverty reduction. In B. Campbell (Ed.), *Mining in Africa: Regulation and development* (chap. 1). London: Pluto Press.

Aryee, B. N. A., Ntibery, B. K., & Atorkui, E. (2003). Trends in the small-scale mining of precious minerals in Ghana: A perspective on its environmental impact. *Journal of Cleaner Production, 11*(2), 131–140.

Aryeetey, E., Laryea, A. D. A., Antwi-Asare, T. O., Baah-Boateng, F. E., Turkson, E. & Ahortor, C. (2004). *Globalisation, employment and poverty reduction: A case study of Ghana*. Geneva: International Labour Office.

Banchirigah, S. M. (2008). Challenges with eradicating illegal mining in Ghana: A perspective from the grassroots. *Resources Policy, 33*(1), 29–38.

Bloch, R., & Owusu, G. (2012). Linkages in Ghana's gold mining industry: Challenging the enclave thesis. *Resources Policy, 37*, 434–442.

Bryceson, D. F., & Jønsson, J. B. (2010). Gold digging careers in rural East Africa: Small-scale miners' livelihood choices. *World Development, 38*(3), 379–392.

Bush, R. (2009). "Soon there will be no one left to take the corpses to the morgue": Accumulation and abjection in Ghana's mining communities. *Resources Policy, 34*(1–2), 57–63.

Ghana Statistical Service. (2007). *Patterns and trends in poverty in Ghana, 1991–2006*. Accra: Ghana Statistical Service.

Gough, K. V. (2010). Continuity and adaptability of home-based enterprises: A longitudinal study from Accra, Ghana. *International Development Planning Review, 32*(1), 45–70.

Gough, K. V., Tipple, A. G., & Napier, M. (2003). Making a living in African cities: The role of home-based enterprises in Accra and Pretoria. *International Planning Studies, 8*(4), 253–277.

Gough, K. V., & Yankson, P. W. K. (2012). Exploring the connections: Mining and urbanisation in Ghana. *Journal of Contemporary African Studies, 30*(4), 651–668.

Heemskerk, M. (2003). Risk attitudes and mitigation among gold miners and others in the Suriname rainforest. *Natural Resources Forum, 27*(4), 267–278.

Hilson, G. (2001). *A contextual review of the Ghanaian small-scale mining industry* (Mining, Minerals and Sustainable Development Report No. 76). London: International Institute for Environment and Development.

Hilson, G., Amankwah, R., & Ofori-Sarpong, G. (2013). Going for gold: Transitional livelihoods in Northern Ghana. *Journal of Modern African Studies, 51*, 109–137.

Hilson, G., & Banchirigah, S. M. (2009). Are alternative livelihood projects alleviating poverty in mining communities? Experiences from Ghana. *Journal of Development Studies, 45*(2), 172–196.

Hilson, G., & Potter, C. (2005). Structural adjustment and subsistence industry: Artisanal gold mining in Ghana. *Development and Change, 36*(1), 103–131.

Langevang, T., & Gough, K. V. (2009). Surviving through movement: The mobility of urban youth in Ghana. *Social and Cultural Geography, 10*(7), 741–756.

Larsen, M. N., Yankson, P., & Fold, N. (2008). Does FDI create linkages in mining? The case of gold mining in Ghana. In E. Rugraff, D. Sánchez-Ancochea, & A. Sumner (Eds), *Transnational corporations and development policy: Critical perspectives* (pp. 247–273). London: Palgrave Macmillan.

Marchetta, F. (2008). *Migration and non farm activities as income diversification strategies: The case of Northern Ghana* (Working Papers – Economics wp2008_16.rdf). Florence: Universita' degli Studi di Firenze, Dipartimento di Scienze per l'Economia e l'Impresa.

Minerals and Mining Act 2006 (Act No. 703). (2006). Retrieved from www.ilo.org/dyn/natlex/natlex_browse.details?p_lang=en&p_country=GHA&p_origin=SUBJECT.

Nyame, F. K., & Blocher, J. (2010). Influence of land tenure practices on artisanal mining activity in Ghana. *Resources Policy, 35*(1) 47–53.

Overå, R. (2008). Mobile traders and mobile phones in Ghana. In J. E. Katz (Ed.), *Handbook of mobile communication studies* (pp. 43–54). Cambridge, MA: MIT Press.

13 Young female entrepreneurs in Uganda

Handicraft production as a livelihood strategy

Sarah Kyejjusa, Katherine V. Gough, and
Søren Bech Pilgaard Kristensen

Introduction

Many young people in Uganda struggle to make a living. Young women in rural areas belong to the most disadvantaged segment of the population; they are frequently less educated, have less access to capital and assets, and have fewer employment opportunities. Moreover, women are still expected to accept primary responsibility for their households and children (Kyomuhendo & McIntosh, 2006) and may even be prevented from earning an income by their husbands for fear of challenging the man's position as head of the household and breadwinner for the family. Handicraft production represents one of the few rural nonfarm activities that is female-dominated and allows many women across sub-Saharan Africa to earn a (mostly) modest income (Rogerson, 2000). It offers opportunities, particularly to women with few resources, to set up a business since it does not require formally acquired skills or substantive start-up capital, who use raw materials which are often readily available in the local area. Furthermore, it is a flexible type of enterprise, which can be combined with household chores and farm work.

Handicraft production is an integral part of Ugandan culture. The types of handicrafts produced include basketry and mats, embroidery and woven products, textiles and hand-loomed products, ceramics and pottery, leather and leather products, wood products, and jewellery and jewellery products (Sector Core Team, 2005). Western Uganda is the prime area of production, though most ethnic groups in Uganda continue to use a range of handicrafts both on a daily basis and for special occasions; as well as having practical uses, many of these objects carry deep cultural connotations. While handicrafts have been made by both men and women in the past, the sector has become increasingly female-dominated (Sector Core Team, 2005). Drawing on qualitative fieldwork conducted in southwestern Uganda, this chapter looks in detail at the role of young female entrepreneurs in the handicraft industry. The study was built around an investigation of the livelihood assets/capitals of the young entrepreneurs, drawing on the sustainable livelihoods framework pioneered by Chambers and Conway (1991) and Scoones (1998). We examine why young women choose to enter this sector and the capitals on which they draw in operating their enterprises. This leads into a discussion of the perception of handicraft production as a desirable business and the likely prospects of the industry, which is sometimes seen as being "old fashioned".

Female entrepreneurship and the handicraft sector

Although there is a considerable gender gap in entrepreneurial activity worldwide – with significantly more men than women being in the process of starting a business or operating new businesses (Kelley, Brush, Greene, & Litovsky, 2011; Minniti, 2010) – in many sub-Saharan African countries, this gap is much narrower. In Uganda, this figure is 32.2% for men compared with 30.5% for women, and the gender gap is reported to be steadily narrowing (Namatovu, Balunywa, Kyejjusa, & Dawa, 2011). Women's businesses, however, tend to be concentrated in certain sectors, are generally smaller with fewer employees, exhibit slower growth, and have lower growth ambitions and limited internationalisation (Hanson, 2009; Kelley et al., 2011). Moreover, in a sub-Saharan African context, a man whose wife works outside the home may be viewed as being unable to control her or provide adequately for his family; hence, many men refuse to allow their wives to run a business (Amine & Staub, 2009).

In a Ugandan context, female entrepreneurship is a relatively new phenomenon born out of war, crisis, and structural adjustment during the 1970s and 1980s (Kyomuhendo & McIntosh, 2006). Today, women in Uganda contribute significantly to the economy, especially through the agricultural sector, textiles and garment industries, and handicrafts. Women tend to enter businesses that have low technical and financial barriers and do not require managerial experience to succeed. These businesses, like elsewhere in sub-Saharan Africa, form part of the informal economy and are best described as micro- or small-scale enterprises, which are often run out of the home (McDade & Spring, 2005). Common problems facing women who want to start or expand their own businesses include poor market information, limited access to capital, and a lack of property to use as collateral (Amine & Staub, 2009).

The handicraft sector worldwide is in a state of flux. Traditionally, rural householders were artisans as well as farmers; during the dry season when there was less work on the farm, men and women would make items such as baskets and mats (Gough & Rigg, 2012). While these products were primarily for their own use or local sale, the rise of tourism has generated a new demand for such products, often in a slightly different form to suit the new market (see Olwig & Gough, 2013, for the example of Bolga baskets in northern Ghana). Consequently, rather than dying out with increasing modernisation, handicraft production is on the rise in many places in conjunction with its commodification and commercialisation (Colin, 2013). Despite the difficulty in collecting reliable statistics, in Kenya, for example, it is reported that the handicraft sector grew by around 70% between 2001 and 2007 (Harris, 2014). Although demand for some handicrafts has risen, the sector as a whole is facing a number of challenges: environmental degradation and climate change are making it increasingly difficult to source raw materials locally, while urban migration, increased competition from industrial products, and changing tastes pose challenges for the viability of the sector (Olwig & Gough, 2013). As Gough and Rigg (2012) show in an Asian context, inputs and even labour are coming from further afield, changing the very essence of the handicraft sector as being locally embedded. There is also a concern that young people are losing interest in the making of handicrafts.

As is the norm elsewhere, in Uganda handicrafts are increasingly being sold to tourists interested in buying "traditional" handmade souvenirs. The handicraft sector, however, is reportedly constrained by a number of challenges, including low production levels, individualised production systems, inadequate supply of raw materials, lack of specialisation, inconsistent product standardisation and quality, and inadequate design skills (Natural Enterprise Development Ltd, 2009). Southwestern Uganda is the country's centre for handicraft production, in particular the districts of Masaka and Mbarara where this fieldwork was undertaken. Most people in the area engage in subsistence agriculture and animal husbandry, and grow coffee as a cash crop, though maize and milk have become nontraditional marketable goods. A study conducted in Mbarara found that 80% of women engage in petty trade and around 30% make handicrafts to sell (Snyder, 2000). In this chapter, we seek to understand the dynamics of the handicraft sector, in particular the opportunities and challenges that young female handicraft producers face. Given that the sector allegedly has difficulties recruiting young people, we are especially interested in understanding who enters the sector and why.

Methodology

In order to locate young female entrepreneurs engaged in handicraft production, the authors contacted local councillors in Masaka and Mbarara who facilitated introductions to relevant respondents. The sample was expanded using the snowballing method. A range of handicraft types was covered in order to explore how young women's experiences vary according to the nature of the product being made. As the aim was to obtain qualitative data to gain an understanding of their motivation, experiences, and business intentions, the young women were interviewed individually by one of the authors using a semi-structured interview guide. The two non-Ugandan researchers were each accompanied by a research assistant who helped with translation where necessary. The interviews were conducted either indoors or outside of where the women operated, which was almost always their home, thus enabling the researchers to observe the products and how they were made. An interview typically lasted 30–40 minutes and a total of 40 interviews were conducted.

Drawing on the livelihoods framework, the young women were asked about resource availability in terms of human capital (the education and skills of the female entrepreneurs as well as any workers employed), natural capital (raw materials used in production), financial capital (sources of finance and income levels), social capital (support from family and friends as well as participation in any networks or associations), and physical capital (equipment used in the business), as well as general information on the type and characteristics of the enterprise (business age, motives, and plans). After presenting a portrait of the young women engaged in handicraft production and the products they make, the analysis is structured around the five key capitals outlined above.

Young female entrepreneurs in handicraft production

The young women interviewed were engaged in producing a range of handicrafts. The most common were baskets and mats (made by half the women), followed by embroidery, crochet, and jewellery (Table 13.1). The products are mainly items to wear or are used as furniture, containers for food, for decoration, or as wedding presents. While most women were engaged in producing only one type of handicraft, some produced a range of products. One such entrepreneurial young woman was Nambusi (aged 24), who knits sweaters and hats, crochets chair-backs, and does embroidery as well as selling vegetables, all from the home. She has become skilled in making a variety of handicrafts as a way of trying to ensure that she always has an income since the demand for her products varies; sweaters, for example, are especially in demand at the start of school terms. Babirye (aged 27) has also diversified her production into several products: she started making baskets and bags from palm leaves, then added making necklaces and earrings from paper after learning from a friend at church that "white people" (tourists) were interested in buying them.

The young women entered the handicraft business both because they perceived an opportunity and out of necessity. For many young women, it is the

Table 13.1 Types of handicrafts produced by women interviewed

Type of handicraft	Number of women
Baskets	15
Mats	14
Embroidery	5
Crochet	5
Jewellery	4
Bags, wallets, and purses	4
Sweaters and blankets	2
Wall hangings	2
Clothes and hats	2
Lids and gourds	1
Total	54
Sample size	*40*

Source: Interviews, 2011.

Note
Several women were undertaking more than one handicraft activity hence the total number of activities exceeds the number of respondents.

only skill they have which they feel they can use to generate an income. Respondents were fairly evenly split between those who had ventured into handicraft production out of interest and those who were engaged in the business because they saw it as the only option available to them. It was common for them to have been inspired by other women to start producing handicrafts. Faith (aged 23) had learned how to decorate gourds (traditional milk containers) from her aunt after moving to live with her, following the death of her parents. The limited capital required to enter the business attracts some young women alongside the perception that there are ready markets for their products.

The young women interviewed were aged 20–32. Most had limited formal education; just over half had not gone beyond primary school and only a few had completed secondary school. Some women cited their limited education as the reason why they were working in the handicraft sector as they felt it was the only work they could do. Many young women were married with small children, and paying school fees and supporting dependants in general was a key incentive for them to engage in handicraft production. There were also a number of disabled young women involved in handicraft production, which is typically seen as a task suited to people with disabilities, given their often limited mobility and education. Nina (aged 22) knits sweaters and blankets using a machine her parents bought for her after attending a vocational school where she attained the skill. She claims that, as a disabled person, handicraft making is her only option to earn money. Nina is not alone in her business, working alongside her nondisabled sister (Figure 13.1).

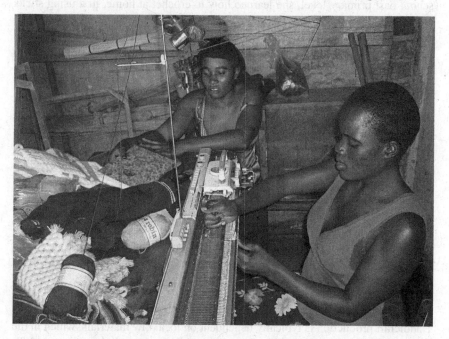

Figure 13.1 Young woman using a knitting machine to make sweaters in Masaka (photo: Søren Bech Pilgaard Kristensen).

Capitals drawn on in handicraft production

Making handicrafts requires drawing on a wide range of capitals. Here, we analyse their importance for young female entrepreneurs, considering human, natural, financial, physical, and social capital in turn.

Human capital

A key aspect of handicraft production is learning the necessary skills. Traditionally, handicraft skills were learned at home and the craftsmanship was handed down from generation to generation. This tradition has waned considerably in Uganda over time (Sector Core Team, 2005). Nowadays, it is more common for young people to learn the necessary skills outside the home, either while in school, through an NGO-run programme (especially jewellery making), or taught by other women in their homes. Basket and mat weaving were skills that some of the young women had learned from their mothers, while others had acquired this skill at school.

Nakawooya (aged 20) learned how to make baskets in handiwork classes at school. As she grew up in a basket-making area, she also helped out at home; by the age of four, she was playing with the materials and from the age of 11 she helped to make baskets. Nambusi (aged 24) had learned the various skills she uses in her handicraft business in a range of settings. After failing to continue in school past primary level, she learned how to crochet at home, first using sticks and subsequently needles. She then learned embroidery by taking a part-time course that lasted a month. After her husband bought her a sweater machine, she paid a woman who lived close by for one week of daily lessons in order to learn the necessary skills. Nambusi hence acquired three different skills in three different ways, all of which she uses in her business.

One surprising aspect was young female entrepreneurs' willingness to teach other young women their skills, despite the fact that this might result in greater competition. As Namulondo (aged 30), who had taught another young woman to make bags and purses out of woven palm leaves explained, she was not worried about creating more competition as "Everyone has their luck." Nambusi had, in turn, trained other women in the community to use a knitting machine for a fee. Whenever she receives many orders, she is able to turn this to her advantage by calling on these women to help her complete the orders. This can be beneficial in a situation where most young entrepreneurs rely on their own human capital to produce handicrafts, only very occasionally employing casual labour, using apprentices, or drawing on the labour of other family members to assist them.

Natural capital

Handicraft production is typically dependent on local raw materials, which in the past could be collected close to where the producers were living (Figure 13.2). For some young female entrepreneurs, this is still the case. Sophia (aged 29)

Figure 13.2 Young woman in Mbarara making baskets from papyrus found in swamps (photo: Søren Bech Pilgaard Kristensen).

weaves mats, which are used to serve cooked bananas (*matoke*). The raw material for her business is papyrus sourced from nearby swamps. She walks to the swamps together with a group of women because it is not safe to go alone due to the risk of being bitten by snakes. The women cut the papyrus and leave it in the swamp to dry for several days before carrying it home on their heads. While the women do not have to pay for their raw materials, collecting them is a time-consuming, tough, and potentially dangerous activity. As the raw materials can only be accessed during the dry season, Sophia's business is seasonal unless she buys papyrus in the market, but this reduces her profit considerably. As other studies have also found, most handicraft producers no longer collect their own raw materials – which, in any case, are often not readily available – but purchase what they need to save time (Gough & Rigg, 2012; Makhado & Kepe, 2006; Olwig & Gough, 2013). Sometimes, market forces put a price on raw materials that were previously free; Sophia reported that people who now live near the swamp have started to try to charge the women to allow them to pass by and collect the papyrus.

As natural materials become harder to access and more expensive, some handicraft producers are turning to artificial raw materials such as plastic and paper. Sourcing these raises new challenges and entails additional costs. The young entrepreneurs making jewellery buy all their raw materials (paper, metal, and superglue). While these young women mostly frequent local markets, Juliet (aged 30) travels to the capital, Kampala, to buy her materials as they are cheaper in the large second-hand market there. For other women, making this journey is not possible due to their responsibilities at home.

Financial capital

Most young entrepreneurs require a degree of financial capital to establish their business. While the limited need for capital to set up in handicraft production was a reason some young women gave for entering the sector, most needed at least some capital to purchase materials and tools in order to start. Some respondents had succeeded in raising their own capital by engaging in other activities, such as farming or making and selling snacks. Family members, however, were by far the most common source of financial capital; for single women, their parents would often provide them with funds, whereas married women tended to obtain start-up capital from their husbands. Asia (aged 25) explained how she ventured into the crochet business because it was a skill she had learned as a girl and she could start the business with UGX10,000[1] provided by her husband. After buying a needle and thread, she spent four hours every afternoon crocheting. Within two months, she was able to accumulate UGX25,000 in profit with which she was able to buy more thread. She expects to be able to continue expanding her business in this way; as she explained, "My products are good because I keep changing fashions and designs of the crochets."

Other women, however, found it frustrating when they could not plough their profits back into their business since there were always family expenses that had to be met. As Namulondo complained, "The money gets tied up with my children's needs." None of the respondents had been able to raise capital from formal financial institutions, mainly due to their lack of collateral. A more common source of finance was women's associations, which operate revolving funds whereby typically all members make a weekly donation that is allocated to one member at the end of the month. Given their often limited financial capital, young female entrepreneurs tend to obtain an advance when they receive an order, also in an attempt to ensure that the customer returns.

Physical capital

The type of physical capital required for handicraft production varies considerably depending on the handicraft. Basket weaving does not require any physical capital in the production process but other activities, such as knitting sweaters, require machines in order to be profitable; Nambusi's knitting machine, which her husband bought her in 2009, had cost UGX330,000. The expense of purchasing such technology was beyond the means of many of the young women and had prohibited them from setting up such, often more profitable, businesses. A universal aspect of physical capital, however, was the need for a place to work.

For most of the young women, this was within their home or in the open space in front of the house which were considered to be adequate and convenient spaces. Some young women had access to shop space to sell their goods and others were hoping to obtain such a space in the future, though for most this was beyond their financial means. Other respondents who sold goods from market stalls would use idle time for production and could store finished and unfinished products there while waiting for another order.

Another key aspect of physical capital is mobile phones, access to which has revolutionised communication in sub-Saharan Africa, where landlines have long been inadequate, and changed ways of doing business (Overå, 2006; Porter, 2012). Almost all the female entrepreneurs used a mobile phone in their businesses to contact customers and suppliers by calling or sending an SMS. A few also used mobile phones to make payments or receive money through "mobile money" schemes. While some young women owned their own phones, others managed by borrowing a phone from another household member. Asia, for example, did not have a mobile phone but her customers could contact her through her spouse. Nambusi used her mobile phone to maintain contact with customers and her network of fellow handicraft producers, mainly through text messages.

Social capital

Social capital is a highly debated concept but it is generally considered to constitute access to resources through social networks and social relations (Lin, 2001; see also Chapter 15). As indicated above, a key aspect of young women's social capital comprises family members on whom they often depend for financial and practical support. For many of the young women interviewed, being a member of a group/association also played an important role in their business activities. For some young women, joining a group had motivated them to start handicraft production in the first place; for others, having access to training, markets, materials, and especially revolving funds through their membership of a group helped them keep their businesses running.

The social aspect of being part of a group – and sometimes being able to work together rather than independently at home – was another key attraction. As Nakayondo explained, "I can accomplish tasks faster when I am sitting together with others. Sometimes when I am alone I feel bored and get lazy." Nambusi had been instrumental in setting up a group a year prior to being interviewed. The group has around 15 members who meet several afternoons a week. Every Sunday, each member of the group pays an amount that is then allocated to one member on a rotational basis to provide capital. Nambusi explained that this capital also enables the group to take on larger orders to make sweaters that they share between them.

Dorothy (aged 20) had also set up a group of six young women who were making jewellery. They had decided to work together so that they could become better known and advertise more widely. After pooling their funds, one of the members travelled to Kampala to buy the materials for the beads, which were cheaper to purchase there. This type of group is built on trust, which can be a challenge to build up. Akankunda (aged 24) explained how a group to which she had belonged disintegrated after some members successfully obtained a grant from the government's "Prosperity for All" programme, while others (herself included) were not able to participate as they did not meet the requirements.

Perceptions and prospects of handicraft production

In order to analyse the future of handicraft production by young women in western Uganda, this section considers how young entrepreneurs perceive themselves and their views of how others perceive them. We then discuss how the young women's businesses are faring and their plans and prospects.

Some of the young people were clearly proud of the work they were doing. They saw themselves as being skilful and creative and, in some instances, playing an important role maintaining a tradition. Sophia claimed to be very proud of her ability to make mats and especially of being able to feed and support her six children as well as pay school fees for those of school age. Other young women were embarrassed though by the handicraft work they were doing, seeing it as a low-status activity for elderly women. This negative self-perception was reinforced by the perceptions and comments of others. Nakawooya explained how people (especially youth) sometimes ridiculed her basket making as something that should be work for old, not young, people. This makes her feel bad and she tries to point out that she has to make baskets as she did not finish school and has nothing better to do. Facing ridicule seems to especially affect those making more traditional products; Nakayondo has never been laughed at for her crochet work, but thought that it was common for youth making mats and baskets to suffer ridicule.

Generally, the profit levels for handicraft production by young women are low. Most of the women interviewed were making monthly profits in the range of UGX10,000 to 50,000, though around one-third were earning below UGX10,000. The more traditional crafts, which are used for practical purposes – primarily mats and baskets – are the least profitable. Products that have a ceremonial purpose or are destined for special occasions, such as weddings, are slightly more profitable. Those crafts that require a specialised technique, such as crocheting and knitting, are among the most profitable; they require more expensive physical capital (yarn and even a knitting machine) and are also subject to higher seasonal demand, such as sweaters for school children at the start of the school calendar, resulting in higher profitability. Despite generally low profit levels, it should not be assumed that all the young female handicraft producers were operating marginal businesses that only "top up" household incomes. Namulondo claimed that her jewellery making brings in more money than her husband's fish selling, with a ratio of about 70:30, despite the fact that she also farms every morning.

Recent years have seen an increase in demand for certain handicrafts and some of the women's businesses are expanding. Many, though, have weak links to the final consumer and, as Snyder (2000) also found, sell their handicrafts through intermediaries instead, thus reducing their profits. Customers who were sold to directly were found by chance encounter or word of mouth and were rarely regular customers, restricting the young women's possibilities for expansion.

Some of the young women had no plans to expand their business, citing primarily lack of time due to other household chores or farming, or insufficient customers, as Makhado and Kepe (2006) also found in a South African context. Most of these women, however, planned to continue running their business as they considered being self-employed in handicraft production preferable to the sort of jobs they could otherwise obtain, such as domestic help where they would have a low wage and risk being exploited. Many of the women, however, spoke enthusiastically about their plans, which included expanding their current businesses, adding new businesses within the handicraft sector, and setting up a new business in an entirely different field. Those who were planning to expand their current business talked about adding new styles, differentiating their products from others, and hoping to acquire premises where they could display and sell their products. Asia told how "As a young person I feel energetic enough to carry on in this business [crocheting]. I wish to grow into a wholesale business and also to get premises to display my goods." Businesses the young women mentioned diversifying into include cattle rearing, second-hand clothing, small-scale retail, restaurants, and hairdressing.

Young women engaged in handicraft production are likely to continue doing so without any official form of support. There are many government and non-government organisations in Uganda designed to promote women's businesses and handicraft production,[2] but they operate in a fragmented and disjointed manner. Most respondents were not aware of programmes related to handicraft production and few had benefited from them in any way. Of greater importance were the many informal organisations set up by the handicraft producers themselves.

Conclusion

Handicraft production by young women in western Uganda is typically based on both informally and formally learned skills and makes use of locally available and/or purchased raw materials. Production is home-based and requires little financial capital input, making it an important entry point into the economy for young women with limited employment alternatives and little formal education. The more traditional products made are baskets and mats, produced with local raw materials either collected by the women themselves or purchased from dealers; newer products include jewellery made from paper beads, all the inputs for which are purchased. Young women making more traditional products sometimes face ridicule for engaging in a task that is seen as being the preserve of old women, while those making more "modern" handicrafts are generally respected for their skills.

These young women combine handicraft production with household chores and farming, and the flexibility of production was a frequently cited motive for entering the trade. The key challenges encountered by the women were access to financial capital and low profitability. Some young women had sought to overcome these by joining associations, which increased their access to financial capital and provided a useful network with other women running similar businesses, thus expanding their social capital. Some had entered the handicraft sector reluctantly as a last resort, due to limited human and financial capital, while others took pride in their skills and appreciated the cultural importance of the trade and the opportunity it provides for income generation. Many had plans to extend their current business and to diversify into other lines of business, either within or beyond the handicraft sector.

The desire for young women in Uganda to continue making handicrafts bodes well for a sector that is sometimes characterised as being in danger of fading out due to lack of interest among youth (Rogerson, 2000). However, a more coordinated government policy and greater institutional collaboration are needed to promote the sector's growth potential. Incorporating vocational training at all levels of the education system would provide young people with the relevant skills for handicraft production. This would be compatible with the recent government initiative of "Skilling Uganda", which aims to change the education system from being largely theoretical to also being skills-based. In particular, government and nongovernment programmes should support young female entrepreneurs' efforts to increase their human capital by equipping them with more technical and entrepreneurship skills. They would also benefit from greater access to financial capital, especially to support their attempts to expand their businesses. Support to the informal associations that play a central role for many young women engaged in handicraft production – increasing their social, human, and financial capital – could greatly increase their potential for production. This is especially important as the handicraft sector provides business opportunities to some of the most vulnerable young women in Uganda.

Notes

1 UGX10,000=US$3.83 at the time of the interviews.
2 These include the Uganda Export Promotion Board, Uganda Women Entrepreneurs Associated Limited, Uganda Small Scale Industries Association, National Organisation of Women Associations in Uganda, National Arts and Crafts Association of Uganda, and Private Sector Foundation Uganda.

References

Amine, L. S., & Staub, K. M. (2009). Women entrepreneurs in sub-Saharan Africa: An institutional theory analysis from a social marketing point of view. *Entrepreneurship and Regional Development, 21*(2), 183–211.

Chambers, R., & Conway, G. R. (1991). *Sustainable rural livelihoods: Practical concepts for the 21st century* (Discussion Paper No. 296). Brighton: Institute of Development Studies.

Colin, F.L. (2013). Commodification of indigenous crafts and reconfiguration of gender identities among the Emberá of eastern Panama. *Gender, Place and Culture, 20*(4), 487–509.

Gough, K. V., & Rigg, J. (2012). Re-territorialising rural handicrafts in Thailand and Vietnam: A view from the margins of the miracle. *Environment and Planning A, 44*(1), 169–186.

Hanson, S. (2009). Changing places through women's entrepreneurship. *Economic Geography, 85*(3), 245–267.

Harris, J. (2014). The messy reality of agglomeration economies in urban informality: Evidence from Nairobi's handicraft industry. *World Development, 61*, 102–113.

Kelley, D. J., Brush, C. G., Greene, P. G., & Litovsky, Y. (2011). *Global entrepreneurship monitor: 2010 women's report*. Babson Park, MA: Babson College.

Kyomuhendo, G. B., & McIntosh, M. K. (2006). *Women, work and domestic virtue in Uganda 1900–2003*. Kampala: Fountain Publishers.

Lin, N. (2001). Building a network theory of social capital. In N. Lin, K. Cook, & R. S. Burt (Eds), *Social capital: Theory and research* (pp. 3–29). New Brunswick, NJ: Transaction Publishers.

Makhado, Z., & Kepe, T. (2006). Crafting a livelihood: Local-level trade in mats and baskets in Pondoland, South Africa. *Development Southern Africa, 23*(4), 497–509.

McDade, B. E., & Spring, A. (2005). The "new generation of African entrepreneurs": Networking to change the climate for business and private sector-led development. *Entrepreneurship and Regional Development, 17*, 17–42.

Minniti, M. (2010). Female entrepreneurship and economic activity. *European Journal of Development Research, 22*, 294–312.

Namatovu, R., Balunywa, W., Kyejjusa, S., & Dawa, S. (2011). *Global entrepreneurship monitor: GEM Uganda 2010 executive report*. Kampala: Makerere University Business School and Danish International Development Agency.

Natural Enterprise Development Ltd. (2009). *Diagnostic study on small and medium forest enterprises in Uganda*. Kampala: Environmental Alert.

Olwig, M. F., & Gough, K. V. (2013). Basket weaving and social weaving: Young Ghanaian artisans' mobilization of resources through mobility in times of climate change. *Geoforum, 45*, 168–177.

Overå, R. (2006). Networks, distance and trust: Telecommunications development and changing trading practices in Ghana. *World Development, 34*(7), 1301–1315.

Porter, G. (2012). Mobile phones, livelihoods and the poor in sub-Saharan Africa: Review and prospect. *Geography Compass, 6*(5), 241–259.

Rogerson, C. M. (2000). Rural handicraft production in the developing world: Policy issues for South Africa. *Agrekon: Agricultural Economics Research, Policy and Practice in Southern Africa, 39*(2), 193–217.

Rogerson, C. M., & Sithole, P. M. (2001). Rural handicraft production in Mpumalanga, South Africa: Organization, problems and support needs. *South African Geographical Journal, 83*(2), 149–158.

Scoones, I. (1998). *Sustainable rural livelihoods: A framework for analysis* (Working Paper No. 72). Brighton: Institute of Development Studies.

Sector Core Team. (2005). *Uganda handicrafts export strategy*. Geneva: International Trade Centre.

Snyder, M. C. (2000). *Women in African economies: From burning sun to boardroom*. Kampala: Fountain Publishers.

14 Employment in the tourism industry

A pathway to entrepreneurship for Ugandan youth

Samuel Dawa and Søren Jeppesen

Introduction

The world tourism industry has grown steadily as a result of the rising incomes of the global middle class and increased awareness of new travel destinations. Given that it is a services-based industry, tourism comprises labour-intensive growth sectors that employ a considerable (and growing) number of people. Accordingly, the industry also holds potential for young people in terms of providing entrepreneurial opportunities and/or employment. The sustained growth of the tourism industry has also meant that an increasing number of travellers now visit destinations in Africa, including Uganda. The country's numerous tourist attractions have enabled it to benefit from the increased number of foreign tourists – which reached close to one million in 2010 (World Travel and Tourism Council, 2011) – as well as from the rapidly growing local demand for services. The industry comprises three main subsectors: hotels and lodging, tour operators, and suppliers of associated goods and services. While the activities of all three subsectors are closely intertwined, this chapter focuses on hotels and lodging and their suppliers.

Although the hotels and lodging subsector offers a number of entrepreneurial opportunities, youth entrepreneurship is limited as most young people are unlikely to have the requisite financial resources to start even a small bed-and-breakfast. Accordingly, the most active participants in the hotels and lodging subsector in Uganda are foreign firms and local investors with substantial means. Still, young entrepreneurs have a wide range of options available in terms of supplying goods and services such as food, beverages, equipment, information and communication technology (ICT), and janitorial and transport services. This chapter seeks to contribute to the literature on youth employment and entrepreneurship by focusing on the nature of youth participation in the tourism industry in Kampala, Uganda's capital city. In this study we seek to gain an insight into how and why young people engage in the sector and their aspirations for the future.

Tourism and hotel industry in Uganda

The tourism sector is Uganda's second largest foreign exchange earner, contributing 7.6% of the country's gross domestic product (Uganda Bureau of Statistics, 2011). It also employs an estimated 447,000 persons (World Travel and Tourism Council, 2011). The number of hotels and lodging places in Uganda has increased over the years from 120 in 1989 (Uganda Bureau of Statistics, 1989) to 3,913 in 2013; the subsector employs a workforce of 15,588 people, of which 90% are aged 18–35 years (UNDP, 2013) and work as receptionists, cleaners, cooks, plumbers, and bookkeepers, etc. Such jobs entail long working hours (often at odd times of the day) and low wages, and can be hard, repetitive, and intensive (Habourne, 1995; Hussain, 2008; Knox, 2011; Mason, Mayhew, Osborne, & Stevens, 2008). The growth of the hotels and lodging subsector in Uganda reflects the country's changing political and economic situation. In the 1960s, the government set up and operated a number of large hotels in tourist havens such as game parks and mountain areas under the umbrella of a parastatal corporation – the "Uganda Hotels". As the economy collapsed, so did the government's ability to maintain tourism infrastructure. In the late 1980s, when the government implemented privatisation policies, many state-owned hotels were sold to private investors. The 2007 Commonwealth Heads of Government summit held in Kampala provided a new impetus with large investments by both government and private individuals in the sector, leading to the establishment of numerous hotels.

Businesses in the hotels and lodging subsector grew by nearly 200% between 2001 and 2011 to 3,876 establishments (Uganda Bureau of Statistics, 2011) employing, on average, eight persons per establishment. Almost half of these are located in the Central Region. Kampala itself offers all types of accommodation from very small, unregistered guesthouses, bed-and-breakfasts, hostels, and lodges to large internationally owned hotel chains such as the Serena and Sheraton. The subsector receives support from an even larger number of associated businesses, including professional services (such as accounting and legal services), functional services (such as laundry and cleaning), and compound maintenance (such as electrical services, gardening and plumbing, and design and security). Given that Kampala has a disproportionately high number of hotels and lodges and, therefore, provides substantial employment and entrepreneurial opportunities for youth, we focus on the two subsectors of the tourism industry in this city.

Methodology

Interviews with hotel and lodging owners and managers, as well as with the youth employed in these enterprises, and the young entrepreneurs who supply the subsector with goods and services were the main form of data collection. Purposive sampling of hotels and lodges was carried out due to lack of reliable statistics on the subsector, which is not categorised based on the international stars system (see, for example, Uganda Bureau of Statistics, 2011), and on the suppliers subsector in Uganda. As Kampala's hotels and lodging places are spread across the city and are characterised by different numbers of employees and employment conditions, a sample was drawn from each of the city's five divisions. In total, 50 establishments were interviewed – 10 in each administrative division (Nakawa, Makindye, Central, Rubaga, and Kawempe). The number of employees in these establishments ranged from four to 550 and the number of rooms varied between five and 220. Semi-structured interviews consisting of both closed and open-ended questions, which lasted 45 to 90 minutes, were conducted by the authors with the help of research assistants. The interviews focused on five main areas: the establishment type and its turnover, categories of clients, and development over time; the type of training activities provided for staff; the number of youth employed in different areas of work and the level of staff turnover; the skills, qualifications, and advantages or disadvantages of employing youth; and the establishment's plans for growth and its opinion of how best to support youth.

Following these interviews, our two assistants traced a number of the suppliers and service providers associated with the hotels and lodging places we had interviewed. In all, we conducted 100 interviews, representing two suppliers of each of the 50 hotels and lodging places surveyed. These interviews focused on the same five areas listed above and with a similar combination of open-ended and closed questions. Each interview lasted between 30 and 60 minutes. Our respondents in this case included 62 young entrepreneurs, which we draw on in the following sections.

Subsequently, in-depth interviews were held with 10 young people: five employed at the hotels surveyed and five young suppliers. We contacted these respondents through five different hotels, each of which provided contact details for one young employee and one youth entrepreneur who supplied the hotel with a particular service. The intention was to obtain the experiences and views of a group of young employees representing a cross-section of functions – a manager, a receptionist, a guest-relations assistant, an IT assistant, a concierge – and a spread of young entrepreneurs engaged in a range of activities – an assortment of food suppliers and garbage collectors. The interviews focused on the young people's employment and life choices including their educational background, family commitments, reasons for the choice of employment, motivation and aspirations in terms of being part of the industry, both at present as well as in the future.

Youth employment and opportunities in the tourism industry

The following subsections present our findings. First, we examine the hotels and lodging subsector and look at the number of young people employed, their plans for growth, their employment conditions and experiences, and their motivation and aspirations for working in this subsector. We then turn to the young entrepreneurs who supply the subsector and present the types of entrepreneurial activities carried out, the gender of the proprietor, the number of employees, their motivation and aspirations for being self-employed versus employed, and their plans for growth.

Hotels and lodging subsector

Of the 50 registered hotels and lodging places surveyed, none was owned by a young person. This is understandable: some establishments had an annual turn-over in excess of US$2 million, which illustrates why young people are not represented in this subsector, given their limited means and minimal opportunities for acquiring the requisite funds. Rogers (aged 29), a hotel manager, explained this clearly, saying that "the hotel business is profitable and I would like to start my own but the funds required are huge". While family, friends, and other personal networks might be able to provide some level of finance for smaller, more modestly priced enterprises, banks and credit institutions are seldom, if at all, willing to lend this kind of money to young entrepreneurs. A large number of youth, however, were employed by the hotels and lodging places; nearly 70% of employees were between 18 and 35 years old with an almost equal proportion of males and females. The reasons managers gave for employing primarily young people varied from "they are more energetic and willing to learn" to "they are easily trainable and learn from their mistakes better" to "it is the 'normal' thing to do and they are easy to find". More than 60% of the owners and managers surveyed said they aspired to grow, through which they expected to employ more young people. This demonstrates the potential of the tourism industry to create youth employment.

When asked about the advantages of being employed in the tourism sector, Rogers stated:

> As a young person starting out in life, you are faced with many costs. Employment in the hospitality sector almost always ensures that at least the cost of food is taken care of since you can have all your meals at work.

As well as the benefits of free meals, Winfred (aged 30), a guest relations manager, observed: "In this sector, you get to meet a lot of people you would not ordinarily meet in other types of employment, enabling you to build networks and improve your confidence in working with different people." George (aged 25), a concierge, and Ruth (aged 24), a receptionist, concurred, saying that a key attraction of working in this sector was the chance to interact with a wide range of people. This view is similar to that put forward by Mkono (2010) in a study on Zimbabwe where respondents identified networking with people from various countries, cultures, and lifestyles and engaging with a diverse workforce as the main attractions of working in the sector.

Some young respondents said they had sought employment in the hotels and lodging subsector because they had trained professionally in some aspect of hospitality. Although the tourism sector has been identified as having the potential to contribute to the economy, it remains short on skilled labour. A number of tertiary institutions, including the government-supported Jinja-based Hotel Training Institute, now offer certificate-level to postgraduate training in hospitality. The specialised nature of this training and the availability of employment opportunities have driven many students to seek employment in the sector. As Rogers explained: "I am qualified in it. This is my field and I like it." Other young employees said they had failed to find employment elsewhere and, therefore, simply settled for the jobs available in the hotels. Some did not intend to stay long, preferring to find employment in sectors that matched their expertise. Ruth gave this account:

> This is my first job since I graduated. I approached the owner of this hotel when the structure was still under construction and he promised he would hire me if I failed everywhere else. I have worked here for almost six months and learnt a lot and that is what I like about this job: there is a lot of training. But I would prefer to work in a sector for which I am qualified.

However, among Ruth's tasks at the hotel are accounts and stocktaking – skills she learnt as part of her diploma in accounts and finance. Thus, she gets the chance to apply some of her skills at her present workplace.

Some young employees indicated that their employment conditions were not always attractive. Ruth voiced her dissatisfaction, saying "Guests harass me especially when I am working at the reception. The night-shift hours are also uncomfortable. I would prefer to go home in the evening." Aaron (aged 26), an IT assistant, shared Ruth's discomfiture: "We are poorly paid; the guests know it and further undermine us by making unrealistic demands." Similar concerns arose when we asked respondents about their employment prospects. Common reasons for not seeking future employment in the sector included the volume of work, the low pay, and the (poor) treatment of staff. George said he would only consider employment in this sector if he failed to find work elsewhere, explaining that his present work was "time-consuming with no holidays or weekends like other sectors". Winfred, who had worked as a guest relations assistant for under a year, was "tired of working in the sector".

Most of the youth respondents interviewed said they did not intend to continue their current employment longer than another two years. While some said they would seek employment in non-tourism firms, others stated that they would venture out on their own and start a business. Some have already done so: one respondent owns a small restaurant, others run businesses not related to the sector. Rogers, who also owns a business, said: "I am currently running a retail store but I hope in the next two years, when I quit employment, to have accumulated enough capital to start a restaurant and health club that I can run myself." Aaron and George both own businesses presently run by their wives while they work at the hotel. George, who owns a restaurant, said:

> I decided to start this business when my former employer decided to terminate a contract to serve food to one of his clients because of a price

disagreement. I felt that, at the price they were offering, I could make a profit, so I approached the company and got the contract.

This shows how attractive entrepreneurship is perceived to be among young people, many of whom have either started a firm already or plan to do so.

Respondents agreed unanimously that, if they were to quit their jobs, they would not seek employment in the hotels and lodging subsector again unless it was as an owner. Ruth stated: "I can only work in this sector if I own the business." Rogers echoed her discontent: "I get frustrated seeing the profit potential of this sector and the bad decision-making by owners. After this employment, I will only participate in this sector as an owner." As mentioned earlier, however, the cost of setting up a hotel or lodge is high, especially for youth who, in most cases, will not have access to resources of that magnitude. Individuals such as Rogers were more optimistic and felt that they could pool their employment and business savings in order to start their own tourism firms. These findings reflect a conflicting scenario: the hotels and lodging subsector requires the energy of youth employees, but the tiring nature of the work, the low wages, long working hours, and poor treatment (and even harassment) lead to a high staff turnover. One-quarter of the hotels reported considerable staff fluctuations. This is line with reports that the hospitality industry has long been plagued by turnover rates traditionally ranging from 60 to 300% (Jones, 2008). In turn, this increases the cost of training new employees.

Some of the hotel managers we interviewed said they were aware of young people who had left their jobs to start enterprises of their own, generally in sectors other than tourism, such as cleaning and janitorial services, retail, hair salons, *boda boda* (transportation on a motorbike) services, and poultry farming. This highlights how aspiring young entrepreneurs use formal employment as a stepping stone to set up their own businesses. Given that it is relatively easy to enter and has a high staff turnover, working in the hotel industry gives many young people the chance to obtain the start-up capital they need. Similarly, Richardson (2010) found that a large percentage of youth employed in the hospitality sector would rather leave and start their own enterprises in another sector.

Suppliers of goods and services to hotels and lodges

Turning to the entrepreneurs that supply goods and services to the hotels and lodging subsector, almost two-thirds of this group are youth, of which more than 80% are male. They engage in a wide range of activities: about one-quarter supply drink and food items (vegetables, meat, poultry, dairy products, and seafood) (see Figure 14.1); one-quarter provide upgrading and/or expansion services (plumbing, fumigation, metal fabrication, carpentry, concrete production, and construction); another quarter provide daily services on the premises (housekeeping, cleaning, gardening, car washing, laundry, salon services, ICT, satellite television services, security, and accountancy) and services including graphics and interior design, entertainment equipment; and the rest deliver services to and

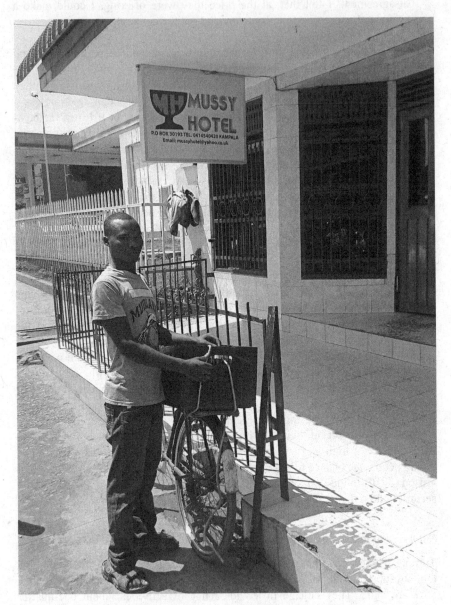

Figure 14.1 Young man supplying drinks to a hotel in Kampala (photo: Samuel Dawa).

from the premises (taxi and *boda boda* transport). Most of the activities mentioned above are undertaken by male entrepreneurs while female entrepreneurs generally only provide housekeeping and laundry services. More than half the young entrepreneurs we surveyed had no employees and very few employed more than five persons, resulting in an additional employment effect of, on average, 0.5 persons per enterprise.

Our findings show that young people choose to engage in the supply subsector for many reasons: they may have based their decision on market research, on their educational background, or simply on a critical incident. Ezekiel (aged 24), an egg trader, was employed as a company sales executive, during which he investigated the viability of supplying hotels with eggs. While the hotel he now supplies is his largest customer, it accounts for only 10% of his business. Bruce (aged 30), on the other hand, was formally employed with the Bank of Uganda before "realising I had a business opportunity to bridge the gap between the farmers and markets". A business administration graduate, he set up a firm and bid for the supply of food items to a public hotel – this contract now constitutes 40% of his sales.

While all the respondents were previously employed, none had ever been employed in the hotels and lodging subsector. Ezekiel simply walked into a hotel and offered to supply the establishment with eggs, to which the management agreed. Raymond (aged 33) attributes the development of his business relationship with his client hotel to the Commonwealth Heads of Government Meeting (CHOGM) held in Kampala in 2007:

> For [the] CHOGM preparations, I was employed by a firm contracted by the Kampala City Council to collect garbage. With the focus being areas where the hotels were located, I realised that the hotels themselves would provide a business opportunity for me.

This shows how entrepreneurial youth become aware of opportunities to set up a business through their formal employment.

According to the young respondents, their biggest incentive to contract with the hospitality sector is that, compared to other firms, their payment is assured. In addition, having a large, known customer adds value to their reputation, which is important to a number of entrepreneurs in terms of gaining other customers. A contract with a well-known hotel signals their trustworthiness to other, smaller clients, and enables young suppliers to expand their customer base. Even in cases where hotels and lodges provide only a small part of their total turnover, for youth entrepreneurs the reputation value of the contract is central to their overall business. As Raymond pointed out: "While the hotel pays little, having such a big client helps you to get other clients." At the same time, all the respondents complained that a key challenge of dealing with hotels was that payments were often delayed. Raymond indicated that although he had collected garbage for a hotel for the last seven years, "While payments are guaranteed, they always come late." Another constraint to suppliers is that, since there are fewer hotels than suppliers, the level of competition is very high.

As far as plans to expand an existing business are concerned, employment is one of the ways in which small-scale entrepreneurship contributes to economic growth (Jeppesen, 2005). Close to 80% of the young entrepreneurs we surveyed said that they hoped to employ more people in the future. The majority hoped to grow by expanding their present business while a few planned to grow by diversifying their business. Only one in five had no plans for growth or expansion. Plans to expand can be manifested in a number of ways. Fred (aged 32), a vegetable vendor, hopes to open another stall in a new market and to start farming on his own. Janet (aged 22) intends to increase her existing capital and stock as a way of expanding her business. Fred (aged 26), a graphics designer, hopes to obtain a loan to purchase modern equipment, while Kato (aged 28), an ICT service provider, plans to acquire bigger premises. In terms of diversification, Mike (aged 33), who operates a carwash, also offers valet parking services to the neighbouring hotel's clients. He plans to branch into the bar, restaurant, and salon business so that "customers do not have to sit idle while we clean their cars". Justus (aged 28), who runs a gardening business, plans to continue this business while starting additional ventures: he finds it natural to run a number of businesses in his desire to emulate other successful businesspersons in his community. Mercy (aged 34), who offers laundry services, also plans to open a restaurant. Interestingly, hardly any respondents plan to expand and occupy the niche held by the hotel or lodging place to which they supply goods or services – mainly because they say they lack the requisite capital. Among those who do not plan to expand their business, Joselene (aged 20) explained her decision as follows: "I prefer to work alone."

In comparing the plans of young men and women, we found that 70% of the male respondents hoped to expand their existing business compared to approximately 50% of the females. Similarly, 82% of the male respondents hoped to employ someone, compared to 64% of their female peers. This supports claims that female-run enterprises tend to be smaller with relatively modest growth plans compared to their male counterparts (see Balunywa et al., 2013; Minniti & Naudé, 2010; Morris, Miyasaki, Watters, & Coombes, 2006; Namatovu, Balunywa, Kyejjusa, & Dawa, 2011). Female entrepreneurs' modest employment intentions compared to their male counterparts might also be explained by a number of sociocultural factors such as family considerations, the absence of role models, their lack of formal business training, poor economic backgrounds, and cultural biases (Morris et al., 2006). While these factors are not unique to the tourism sector, the types of activities suppliers undertake, where they move from customer to customer and do not work from home, seem to favour male entrepreneurs over their female counterparts.

Conclusion

The tourism industry is a growing, labour-intensive sector in Uganda. While the hotels and lodging subsector needs the vitality and energy that comes with employing youth, the young persons we interviewed expressed a range of reservations regarding employment in this subsector. Although youth constitute the majority of employees in this sector, their participation with regard to hotel or lodging ownership is limited mainly because of the large capital investment required. Young entrepreneurs use the sector as a way of earning capital in order to set up their own businesses. The tourism sector has several advantages in this regard: young people who eat at work can save much of their (low) salary; coming into contact with numerous people helps them build up their social networks; and they are in a position to perceive opportunities for new businesses supplying and servicing the hotels. Many more opportunities for young people to set up a business exist in the suppliers subsector, given the very wide range of goods and services delivered to hotels and lodges. Here, youth constitute the majority of entrepreneurs and are predominantly male. Young entrepreneurs tend to employ a limited number of persons, but most do have plans for growth. Of these, the majority plan to expand their current business while a minority hope to open up new lines of business. Although our study found that youth were not represented as hotel or lodging owners, some young employees expressed such aspirations, pending access to sufficient funds.

Policies that provide incentives to support the current growth trend in the tourism industry – whether in terms of increasing the number of tourists or expanding the number of hotels and lodging places – would benefit youth employment and ensure extended entrepreneurial opportunities for other young persons engaged in supplying services and goods to hotels and lodges. However, ensuring that employment in this sector remains an attractive prospect for young people is more challenging and will need further study. At the moment, young employees seem to prefer self-employment to working at a hotel or lodge. Similarly, further investigations are needed to detect how best to support youth entrepreneurs in expanding their businesses and employing additional labour. Furthermore, this chapter has highlighted how interconnected the formal and informal sectors are. Many young people become entrepreneurs initially by working in the formal sector in order to obtain the necessary start-up capital, build up their social networks, acquire skills, and gain good ideas for new businesses. The tourism industry, despite its high staff turnover, provides young people with much-needed experience on which to draw when running their enterprises. Informal micro-enterprises, many of which are run by young people, provide the hotels and lodging subsector with supplies and services. Therefore, although this subsector does not provide many openings for young entrepreneurs, it plays an indirect but important role in stimulating youth entrepreneurship.

References

Balunywa, W., Rosa, P., Dawa, S., Namatovu, R., Kyejjusa, S., & Ntamu, D. (2013). *Global entrepreneurship monitor: GEM Uganda 2012 executive report*. Kampala: Makerere University Business School and International Development Research Centre.
de Silva, L. R., & Kodithuwakku, S. S. (2011). Pluriactivity, entrepreneurship and socio-economic success of farming households. In G. A. Alsos, S. Carter, E. Ljunggren, & F. Welter (Eds), *The handbook of research on entrepreneurship in agriculture and rural development* (chap. 3). Cheltenham: Edward Elgar.
Habourne, D. (1995). Issues in hospitality and catering. *Management Development Review, 8*(1), 37–40.

Hussain, H. (2008). A-hunting we will go. *Caterer & Hotelkeeper, 198*(4510), 56.

Jeppesen, S. (2005). Enhancing competitiveness and securing equitable development: Can small, micro, and medium-sized enterprises (SMEs) do the trick? *Development in Practice, 15*(3–4), 463–474.

Jones, J. (2008). A unique formula for success. *Lodging Hospitality, 64*(15), 60–64.

Knox, A. (2011). "Upstairs, downstairs": An analysis of low-paid work in Australian hotels. *Labour and Industry, 21*(3), 573–594.

Mason, G., Mayhew, K., Osborne, M., & Stevens, P. (2008). Low pay, labour market institutions, and job quality in the United Kingdom. In C. Lloyd, G. Mason, & K. Mayhew (Eds), *Low-wage work in the United Kingdom* (pp. 41–95). New York, NY: Russell Sage Foundation.

Minniti, M., & Naudé, W. (2010). What do we know about the patterns and determinants of female entrepreneurship across countries? *European Journal of Development Research, 22*(3), 277–293.

Mkono, M. (2010). In defence of hospitality careers: Perspectives of Zimbabwean hotel managers. *International Journal of Contemporary Hospitality Management, 22*(6), 858–870.

Morris, M. H., Miyasaki, N. N., Watters, C. E., & Coombes, S. M. (2006). The dilemma of growth: Understanding venture size choices of women entrepreneurs. *Journal of Small Business Management, 44*(2), 221–244.

Namatovu, R., Balunywa, W., Kyejjusa, S., & Dawa, S. (2011). *Global entrepreneurship monitor: GEM Uganda 2010 executive report.* Kampala: Makerere University Business School and Danish International Development Agency.

Richardson, S. (2010). Understanding Generation Y's attitudes towards a career in the industry. In P. Benckendorff, G. Moscardo, & D. Pendergast (Eds), *Tourism and Generation Y* (pp. 131–142). Wallingford: CAB International.

Uganda Bureau of Statistics. (1989). *Census of business establishments 1989.* Kampala: UNDP.

Uganda Bureau of Statistics. (2011). *Report on the census of business establishments, 2010/11.* Kampala: World Travel and Tourism Council.

United Nations Development Programme. (2013). *Uganda domestic tourism survey report, 2013.* Kampala: Uganda Bureau of Statistics.

World Travel and Tourism Council. (2011). *Travel and tourism economic impact 2011: Uganda.* London: Uganda Bureau of Statistics.

Concluding comments to Part IV

Katherine V. Gough and Thilde Langevang

The chapters in this part of the book, which studied young entrepreneurs in four differing sectors of the economy, reveal a wealth of details highlighting the similarities and differences young people face in running businesses between and even within certain sectors. These are highlighted here focusing on how the respective sectors are viewed by the young entrepreneurs and the population as a whole, why they enter the sector, how the sectors are gendered, and how long young entrepreneurs tend to remain within a sector.

Views of the various activities young entrepreneurs engage in vary by sector and within sectors depending on the specific tasks the young people undertake. Broadly speaking, establishing a business within the mobile telephony and tourism sectors is viewed as being positive by both the young entrepreneurs and others. Some young entrepreneurs, especially within the mobile telephony sector, are passionate about their businesses. Working in illegal small-scale mining and in the traditional handicraft sector, however, both carry a degree of stigmatism, though in differing ways. In the case of mining, the work is viewed as being highly dangerous and detrimental to the environment, and many young people who enter the sector are castigated for just wanting to make quick money (Hilson & Osei, 2014). However, the financial benefits to be gained from entering the sector are gradually being recognised and small-scale mining is not as stigmatised as it has been, especially among the young people themselves, though it continues to be problematised by the media and policy makers. The handicraft sector, on the other hand, is seen by many youth as being an old-fashioned backward sector that is only to be entered as a last resort by young people (especially young women) who have dropped out of school. Young women engaged in the more traditional handicraft activities, such as weaving baskets and mats, reported facing ridicule and some felt embarrassed about their work. Others, however, managed to retain a degree of pride, emphasising the skilled nature of their work. One reason for the lack of respect afforded handicraft production is that, contrary to the other sectors studied, it is a business operated from the home. Such home-based enterprises tend to be run by women and are often seen as not being important or even "real" businesses by those operating them and by others (Gough, 2010).

The reasons why young people establish certain businesses tends to cut across the sectors. Young people often engage in the more basic tasks, such as selling airtime for mobile phones or working in a menial job in mining, in order to generate some capital either to support their continued education or to generate capital to move on to better businesses within the same sector or switch into other sectors. Similarly, young people who are employed within the tourism sector also tend to see their work as temporary and a means of generating capital to set up a business, often drawing on contacts they have established to become an entrepreneur supplying hotels and lodges. Consequently there is a frequent turnover of young people in these business types. Young people within the mobile telephony business are especially keen to remain in the sector and those with the necessary resources rise up through the ranks, as is also the case for more educated youth within the small-scale mining sector. Handicraft production differs in some respects to the other sectors studied as many young women remain doing the same activity within the sector, though it is common for the more resourceful young women to both add additional crafts to their repertoire and, where possible, invest in machinery. Importantly, across all four sectors, neither the activities undertaken nor the young entrepreneurs doing them are static as is commonly found in a wide range of business types (Jones, 2010; Langevang, Namatovu & Dawa, 2012).

Income-generating activities within sub-Saharan Africa tend to be gendered but as Overå (2007) has shown, the informalisation of the labour market has resulted in the blurring of some of these gender boundaries. Within all four sectors discussed, there is range of different business types that tend to be gendered and fall into a hierarchy, with women typically running the least attractive businesses. For example, although women are found far more often than would be expected operating businesses in small-scale mining settlements, young women working within mining itself tend to have the menial jobs, the more prestigious and profitable ones being the (almost) sole preserve of men. Similarly, within the mobile telephony sector it is the educated men who are found working up the chain running the larger businesses. Although the handicraft sector is dominated by women, there is still a hierarchy in the tasks undertaken with the more resourceful young women who can draw on a wider range of capitals making the more "modern" and more profitable handicrafts.

As other chapters have shown, it is often young people who are the most entrepreneurial being quick to spot a new opportunity. This was also shown to be the case within these sectoral studies and is especially apparent in the newly established mining settlement in Ghana where a whole raft of businesses has been set up primarily by young people to cater for the needs and newly generated money of the migrant miners. Contrary to the other sectors studied, most residents in the mining settlement only live there for part of the year hence their businesses are only short-lived and migration is a central part of their livelihood strategies (Banchirigah & Hilson, 2010). This contrasts quite sharply with handicraft production where many of the young entrepreneurs are women with young families whose mobility is restricted and they appear likely to continue operating the same business for lengthy periods of time. While some young people in all of the sectors studied, though especially in the mobile telephony sector, envisage remaining in the same sector (though not necessarily in the same business), for others their enterprise is just seen as offering them the chance to earn some money in the short term before moving on to other activities.

These sector-specific chapters thus support the claims made in preceding chapters for the heterogeneity of experiences of young entrepreneurs in sub-Saharan Africa. They highlight the need for policy makers to consider differences not only between sectors of the economy but also within sectors, and the ways in which opportunities vary due to the social characteristics of the young entrepreneurs.

References

Banchirigah, S. M., & Hilson, G. (2010). De-agrarianization, re-agrarianization and local economic development: Re-orientating livelihoods in African artisanal mining communities. *Policy Sciences, 43*(2), 157–180.

Gough, K. V. (2010). Continuity and adaptability of home-based enterprises: A longitudinal study from Accra, Ghana. *International Development Planning Review, 32*(1), 45–70.

Hilson, G., & Osei, L. (2014). Tackling youth unemployment in sub-Saharan Africa: Is there a role for artisanal and small-scale mining? *Futures, 62*, 83–94.

Jones, J. L. (2010). "Nothing is straight in Zimbabwe": The rise of the Kukiya-kiya economy 2000–2008. *Journal of Southern African Studies, 36*(2), 285–299.

Langevang, T., Namatovu, R., & Dawa, S. (2012). Beyond necessity and opportunity entrepreneurship: Motivations and aspirations of young entrepreneurs in Uganda. *International Development Planning Review, 34*(4), 242–252.

Overå, R. (2007). When men do women's work: Structural adjustment, unemployment and changing gender relations in the informal economy of Accra, Ghana. *Journal of Modern African Studies, 45*, 539–563.

Part V

Stimulating youth entrepreneurship

Introduction to Part V

Katherine V. Gough and Thilde Langevang

Young entrepreneurs need a variety of resources and support from a range of actors and organisations to start and expand their businesses. While common discourses of entrepreneurship depict entrepreneurs as heroic individuals who operate in isolation from the surrounding society, there is increased acknowledgement that entrepreneurs are dependent on social relations and rely on supportive institutional environments and assistance from a range of actors and institutions (Langevang, Namatovu, & Dawa, 2012). Moreover, evidence shows the need to differentiate between the youth and adult populations when devising support mechanisms as young entrepreneurs face particular constraints and incentives (Schoof, 2006). This part of the book takes a close look at particular support structures and schemes initiated to stimulate youth entrepreneurship in specific contexts. The chapters scrutinise three vital stimuli to entrepreneurship – social networks, microfinance, and education – which are provided by a range of actors including family, friends, microfinance institutions (MFIs), and universities.

Social network approaches to entrepreneurship have gained prominence in recognition that entrepreneurs are not isolated individuals but are tied to other actors through their social relationships. Relationships are the key medium through which entrepreneurs gain access to a variety of resources held by other actors such as information, advice, emotional support, and finance (Kuada, 2009; Turner & Nguyen, 2005). The concept of "social capital" has been embraced by development researchers and agencies alike to highlight how social relations can influence access to resources (Woolcock & Narayan, 2000). The first chapter in this part examines the role of social capital for young entrepreneurs operating in marketplaces in Lusaka. In view of the low level of governmental support to entrepreneurs in Zambia and the informal nature of their businesses, the chapter examines how and with what effect young entrepreneurs mobilise resources through their social networks and, in particular, their use of social capital to access information, finance, skills, and customers.

Access to finance is one of the most important facilitators for entrepreneurship and is one of the greatest constraints to youth entrepreneurship. Young people face particular constraints to accessing finance as they often lack securities and credibility, business experience, and track records (Schoof, 2006). In sub-Saharan Africa, the informal character of most enterprises amplifies these impediments. MFIs have mushroomed in both rural and urban areas of sub-Saharan Africa in recent years to provide financial services to entrepreneurs who operate in the informal sector and do not have access to commercial banks. Many of these MFIs, however, are shying away from providing financial services to young people whom they typically consider to be too risky clients (Shah et al., 2010). The second chapter in this part focuses on three MFIs in Kampala, which have successfully adopted particular financing approaches for young people. It examines the characteristics of these approaches and the factors that account for their successful development focusing in particular on the role innovation and an entrepreneurial environment play within the MFIs.

Much formal education in sub-Saharan Africa has been criticised for equipping young people with the skills to become "job seekers" rather than "job creators". In response to the broad acknowledgement that it is vital to assist young people to develop entrepreneurship skills, attitudes, and behaviours, there has been an increased focus on entrepreneurship education promoted by donor organisations, nongovernmental organisations, and governments (DeJaeghere & Baxter, 2014). In the light of increasing youth unemployment, educational institutions from the primary level upwards are increasingly being pressurised by policy makers and students to offer entrepreneurship courses. The final chapter of this part, and of the book as a whole, explores entrepreneurship education at the university level in Uganda. It examines the experiences of entrepreneurship graduates from two universities in Kampala, revealing who studies entrepreneurship education and why, and examines how their studies have impacted on their entrepreneurial intentions.

References

DeJaeghere, J., & Baxter, A. (2014). Entrepreneurship education for youth in sub-Saharan Africa: A capabilities approach as an alternative framework to neoliberalism's individualizing risks. *Progress in Development Studies, 14*(1), 61–76.

Kuada, J. (2009). Gender, social networks, and entrepreneurship in Ghana. *Journal of African Business, 10*(1), 85–103.

Langevang, T., Namatovu, R., & Dawa, S. (2012). Beyond necessity and opportunity entrepreneurship: Motivations and aspirations of young entrepreneurs in Uganda. *International Development Planning Review, 34*(4), 242–252.

Schoof, U. (2006). *Stimulating youth entrepreneurship: Barriers and incentives to enterprise start-ups by young people* (No. 388157). Geneva: International Labour Organization.

Shah, A., Mohammed, R., Saraf, N., & Nigam, R. (2010). *Microfinance services for youth in the sub-Saharan African region.* Coventry: Warwick Social Enterprise and Microfinance Society.

Turner, S., & Nguyen, P. A. (2005). Young entrepreneurs, social capital and Doi Moi in Hanoi, Vietnam, *Urban Studies, 42*(10), 1693–1710.

Woolcock, M., & Narayan, D. (2000). Social capital: Implications for development theory, research, and policy. *The World Bank Research Observer, 15*(2), 225–249.

15 Social capital among young entrepreneurs in Zambia

Moonga H. Mumba

Introduction

In Zambia, entrepreneurship has become important for integrating young people into the labour market (Chigunta, 2007). A major challenge, however, is that, despite the high incidence of business start-ups in the country (Amorós & Bosma, 2014), entrepreneurs are offered few resources through official channels (Hansen, 2010). With limited official support, social capital plays a vital role in helping people gain access to resources (International Labour Office, 2002). A basic assumption of social capital is that a person's family, friends, and associates constitute an important asset – one that can be called on in a crisis, enjoyed for its own sake, and leveraged for material gain (Woolcock & Narayan, 2000: 225). This chapter analyses the role of social capital among young entrepreneurs in mobilising various resources to support their businesses in the informal economy in Lusaka's Mtendere and New Soweto markets.

The chapter is divided into four sections. The next section clarifies the concept of social capital. This is followed by an analysis of the use of social capital to access business resources such as information, finance, skills, and customers, which young entrepreneurs have identified as key aspects of starting and running a business. The chapter concludes by arguing that, while social capital helps young entrepreneurs achieve certain ends in pursuit of business, it cannot be solely relied on as a basis for entrepreneurship development among the country's youth.

Social capital in Zambia

Although social capital does not have a universally accepted definition, most advocates of the concept agree that it incorporates the advantages resulting from social connections with other people (Deakins & Freel, 2009). According to Narayan and Cassidy (2001: 60), "to possess social capital, a person must be related to other people, and it is those others, not himself [or herself], who are the actual source of advantage". Social capital focuses on the resources embedded in one's social networks and how access to and use of such resources benefit the individual's actions (Tanga, 2009: 447). Bourdieu, one of the key authorities on social capital, defines it as "the sum of the resources, actual or virtual, that accrue to an individual or a group by virtue of possessing a durable network of more or less institutionalised relationships of mutual acquaintance and recognition" (cited in Burt, 2001: 32). Coleman, another source often cited, defines social capital as "a function of social structure producing advantage" (cited in Burt, 2001: 32). The aphorism "It is not what you know, it is who you know" thus sums up the conventional wisdom on social capital (Woolcock & Narayan, 2000: 225). Like other forms of capital, social capital makes it possible for people to achieve certain ends that might otherwise not be attainable (Burt, 2001: 32).

From an entrepreneurial perspective, Granovetter (1985) has inspired numerous studies in analysing the role of social capital (van der Leij, 2006). He argues that economic actions are embedded in networks of personal relationships and economic goals are typically accompanied by noneconomic goals that are related to the social context. Granovetter also stresses the unique role of concrete personal relations and their structures in the daily work and accomplishments of all sorts of economic actors (Granovetter, 1992). He sees social embeddedness as encompassing a mix of both strong and weak ties that fall more or less within the radius of an individual entrepreneur (Granovetter, 1985: 490). Strong ties generally include intra-group arrangements that are heavily based on the social relationship or personal interaction between actors such as one's family (kin), close friends, church, or neighbours. Weak ties, on the other hand, are associated with single-strand relationships that are limited in terms of intensity, frequency, and reciprocity; these can include acquaintances or friends of friends, former classmates, and co-workers. Weak ties provide strategic advantages in accessing resources beyond one's egocentric networks (Granovetter, 1973). Granovetter's notion of social embeddedness in social relations is adopted as the analytical framework for this study.

Social capital plays an important role among Zambians. It is embedded in traditional ways of life where support is easily given to others, especially among relatives and community members. In the context of a market economy, social capital emerges both as a result of vulnerability and a strategy against vulnerability. People use social capital, therefore, to support one another and reduce their own vulnerability (Mulenga, 2000).

Among many urban Zambians, survival strategies can be linked partly to the existence of an informal economy with markets playing a central role. Over the years, these have become centres of political, social, economic, and cultural life for urban low-income groups in Zambia (Hansen, 2010). These markets provide an income to traders and many urban dwellers have raised their families on the market income they earn. Markets also provide a variety of goods and services required by a wide range of people, especially the poor. Goods are repackaged and sold in the quantities required by the purchaser, making them affordable to a wide range of customers. Markets also stock certain indigenous foods that are desired by not only low-income residents but also some rich people. Bargaining for better prices is part of the market culture, with customers invariably negotiating for lower prices. Finally, markets are also spaces for social interaction in cities where traders interact with their customers on a regular basis and develop relationships that allow people to obtain goods on credit (Hansen, 2010; Nchito, 2006).

Methodology

This research was undertaken as part of doctoral work looking at the role of informal support networks in sustaining self-employment among youth in Zambia. The primary target population for the study included young males and females aged between 15 and 35 years who were self-employed in the informal economy of Lusaka city. Given that markets have become centres of political, social, economic, and cultural life for low-income urban dwellers (Hansen, 2010), it was decided to focus on young entrepreneurs running businesses in such marketplaces. Lusaka has approximately 57 designated markets acknowledged by the Lusaka City Council and numerous others that are not recognised.

Two different markets, New Soweto and Mtendere, were selected for the study as both have a large proportion of youth engaging in a wide range of business activities. Soweto market is Lusaka's largest market and is centrally located in the city. New Soweto market, as it is now known, is an ultra-modern market run by Lusaka City Council. It is equipped with state-of-the-art infrastructure, although makeshift structures are also becoming the norm following a presidential directive that allows vendors to trade freely anywhere.

Mtendere market, on the other hand, is one of the largest and oldest markets in Mtendere compound – an unplanned residential township in Lusaka located about 8 km east of Lusaka city centre. Unlike New Soweto, Mtendere is run by its members through a market cooperative (although there is a planned move by the Patriotic Front government to transfer all markets to local councils and municipalities under the Ministry of Local Government). The market comprises various business activities in makeshift structures, typical of most markets in densely populated residential areas in Lusaka (Hansen, 2010).

Fieldwork data collection combined both quantitative and qualitative methods and was conducted between December 2011 and August 2012. This chapter builds on the qualitative data collected, which consisted of face-to-face in-depth interviews with 10 specially selected young entrepreneurs operating a range of businesses – five in each of the selected markets (see Table 15.1). Each participant was visited several times to learn more about their lives and to confirm initial impressions. The frequent visits helped build trust between the young entrepreneurs and the researcher, enhancing the quality of the data collected.

The initial interviews began with a self-introduction and preliminary questions to ascertain business ownership among the selected interviewees. Each interviewee was asked the broad question, "Tell me how you began this business?" and subsequently asked about their business trajectory and how they had mobilised various resources related to their business. The interviews were conducted at participants' business premises in order to observe them in action and learn more about their businesses. Some interviewees also agreed to let me speak

Table 15.1 Profiles of young entrepreneurs

Entrepreneur	Business activity	Location	Education and qualifications
Innocent, male, 24 years old	Metal fabrication	Mtendere	Grade 9
Mary, female, 30 years old	Cosmetics and hairdressing	Mtendere	Grade 9
Mirriam, female, 26 years old	Vegetables	Mtendere	Grade 9
Phillip, male, 35 years old	ICT services	Mtendere	Grade 12, certificate in accounts
Tobias, male, 32 years old	Hardware goods	Mtendere	Grade 8
Anna, female, 25 years old	Restaurant	New Soweto	Grade 11
Chiti, female, 29 years old	Fashion design	New Soweto	Grade 12, certificate in fashion design
Jackson, male, 29 years old	Shoemaking and repair	New Soweto	Grade 7
Mutale, male, 34 years old	Automobile spares	New Soweto	Grade 12
Thomas, male, 29 years old	Fruits and fresh vegetables	New Soweto	Grade 12, certificate in quality control

to their "visitors" in order to learn more about their relationship with the interviewee. Besides recording the interviews and taking notes, photographs were taken after having obtained the necessary consent.

Following the interviews, the raw field notes were typed up and edited, and the tape recordings transcribed. The interviews were then coded line by line to highlight themes relating to each young person's experience of how different structures had influenced their business. Selected themes, phrases, and sentences were then analysed to identify emerging commonalities and differences in the experience of self-employed youth in mobilising various resources for their businesses.

Acquisition of business-related information

Information is a critical resource for anyone interested in starting a business: the type of business to be established, how it will be launched, and where it will be located, etc. Young entrepreneurs indicated they had drawn on strong as well as weak ties for business ideas and information. Strong ties included talking to friends and, in some cases, family members. Casual conversations with friends had helped elicit business information, as Tobias and Thomas both explained:

It was through casual talks with friends that opened up my mind to some ideas related to running a business.

(Tobias)

The idea [for this line of business] came through chatting with a close friend from another country, who used to live here [Lusaka]. My friend asked [me] to accompany her to buy butternuts at a nearby marketplace after a church service. Surprisingly, butternuts happened to be among the produce that the company I used to work for was exporting to Europe but I had no idea how profitable they were. I did not even know that they had some demand locally. She even assured me that if I had a reliable source and started a business of selling butternuts, I will be assured of the market through her connections. That was how I quit employment and started business.

(Thomas)

While strong ties emerge as a source of information, we also see that, in the case of Thomas, they provide a mechanism that invokes links to weak ties, i.e. his trusted friend's assurance that she will refer him to further contacts. Thomas considered this information reliable and even made up his mind to resign and begin a business – a decision he said he did not regret. Granovetter (1985: 490) argues that information and support gained through strong ties offer multiple benefits as it is: cheap, richer, more detailed, and more accurate (hence, more trustworthy) information, usually from a continuing relationship and so in economic terms more reliable.

Tobias said he had continued to rely extensively on his friends for business information and ideas despite feeling more established in business. He stressed the importance of sharing information with his friends on how he had learnt to vary the commodities sold in different seasons – a factor to which he attributed his success. Tobias said he varied his merchandise based on what he termed the "demands of the season". As he explained:

> Even this time when my business is more established, there are times when business gets tough, especially around the middle of the month when usually all of us are affected. During such hard times, instead of sitting or standing alone by our own stands [shop], we sit together with friends at one shop, playing drafts, chess, or just chatting. In the process, we engage in discussions of all kinds. But sometimes, you find that someone touches on a certain idea that you find appealing to your business. From there, you may want to find out more or even just try implementing such an idea.... I have found trying some of those ideas in my business working very well for me. If it were not because of sharing information with my friends, I would not have been in business today. In fact, nowadays, a day cannot pass without me touching money [meaning, without selling something]. Meanwhile, previously, I would at times open my shop in the morning and only close in the evening without selling anything at all.

These findings on how the self-employed access business-related information seem to be consistent with those of Roever (2005: 34), who shows how, in Peru, collective occupation of a common urban space among informal economy operators provides a setting in which individuals develop networks, have the opportunity to discuss common problems, and share and exchange information.

Regular customers also serve as another important channel for accessing business-related information. For some young entrepreneurs, people who were initially ordinary customers had since become friends who go directly to their respective shops any time they want items related to their own businesses. Anna, Thomas, and Mutale, for instance, had worked for other people's businesses or helped relatives before starting their own business. The experience had given them the opportunity to interact with customers and acquire business ideas and information that had later benefited their own ventures. Interaction with customers was thus not strictly confined to the time they spent running their own business, but also included their time as employees of other people. Their relationships with some clients had, however, continued once they had established their own business.

Access to financial resources

Access to finance is a key resource required for business. However, lack of finance poses the most intractable challenge confronting entrepreneurs in Zambia. The cost of borrowing money from formal lending institutions is generally high (above 15%). Furthermore, commercial banks do not usually extend credit facilities to people working in the informal economy, especially those without collateral (Mwenechanya, 2007). Given this scenario, entrepreneurs in the informal economy tend to look elsewhere for financial support. This study revealed young entrepreneurs' heavy use of strong ties such as the family to mobilise financial resources for business. As Mutale indicated:

> Even if it has not been an easy struggle to get financial support, my family cannot throw me anywhere [meaning, cannot let me suffer]. They know that somehow I have to stand on my own and when I do so through business, they will also benefit. In fact, this is what I am doing right now. After getting financial support from my brother, I take care of my own siblings in paying for their school.

Family support is considered an obligation in the African set-up and is exercised through instruments of norms of exchange or reciprocity. People generally trust that those they have supported will help other family members in turn once they have grown up or are able to stand on their own feet. Tanga (2009) in Lesotho has also established that, in the absence of reliable and established financial support structures, the family is a primary source of care and support for needy members, especially the elderly and children. Some young entrepreneurs, however, felt that financial support from the family was usually inadequate to meet the expressed financial requirements of their business. Economic pressure seems to have affected and changed family systems, especially the extended family system in Africa (Tanga, 2009).

For some young entrepreneurs, obtaining financial support even within their own family circles (strong ties) was difficult. In some instances, this was not due to the family's inability to provide support; rather, it stemmed from fear of competition, jealousy, doubting the ability of the member seeking help to successfully run a business and, in some cases, not approving of the type of venture. "One has to '*be hard*' to get financial help even from within family circles", as Mutale put it, and Anna, Chiti, and Innocent expressed similar sentiments. Chiti, whose mother had exposed her to her current business, highlighted:

> My mother initially said she was not going to help me to get established with my choice [tailoring and fashion design] even though she had the means to do so. She told me that my results were good and I should settle for a better career like journalism or anything more decent for a grade 12 with good results. She even banned me from going to her shop [where she used to do her tailoring business] as a way of discouraging me so that I would settle for another occupation. But I had to insist. It took my mother a long time to accept my decision but later, though reluctantly, she gave in to my desires. Besides money, she even gave me a sewing machine, which was what I wanted the most to stand on my own in business.

Similarly, Anna (a restaurant owner) had no easy way of obtaining financial support from her family. Her own mother, despite having just received her retirement benefits as a teacher, refused to assist her. As Anna explained:

> My mother and aunt discouraged me saying that, being on my own in business as a young lady, I would just end up being abused by men and have another pregnancy, if not contracting HIV/AIDS and die. Instead of getting their support, they encouraged me to look for another job. So I had to go it alone in the initial phase.

In Mutale's case, he had to plead with his elder brother for a long time before the latter agreed to help him financially. Even after he had convinced his brother to help him, the arrangement his brother suggested further delayed Mutale in executing his business plans:

> It took me a long time to convince my brother to give me money to start a business of my own. When my brother finally agreed, instead of giving me hard cash, he said we should continue working together and he gave me one day every week to collect whatever amount of money we realised from business. This was very discouraging to me as sometimes nothing was realised on *my* day and it took me a long time to build up my capital base. At some point I even thought of abandoning my idea of standing on my own in business.

Such revelations appear to be consistent with Locke and Verschoor's (2008) findings, which indicate that trajectories for young people in Zambia are characterised by delays, interruptions, incompletion, false starts, chance opportunities, reversals, and adverse events and interaction.

Besides family support, two other channels used by some young entrepreneurs to obtain money – mostly for business expansion – are *chilimba* (an informal revolving fund) and *kaloba* (loans from illegal moneylenders). Although *chilimba* is fairly popular among young entrepreneurs, things do not always go well for everyone. Nonetheless, members reported finding ways to help each other, depending on the circumstances that had led a person to fail to meet her or his debt obligations. Such mechanisms included swapping turns to receive money, paying for one another and, in some instances, having the debt cancelled.

Kaloba was identified as the easiest way to obtain money quickly for business purposes. However, for young entrepreneurs, it is usually a last resort because it is said to be risky and too expensive. As Tobias explained:

> More often you are required to pay back more than double the amount borrowed even in a period as short as one week. Failure to do so results in confiscation of your merchandise or some valuables and sometimes people end up closing down business.

Almost all the young entrepreneurs shared this view. One young entrepreneur, formerly a lender, observed that the risks to both lender and borrower were too high.

Acquisition of business skills

Skills help promote entrepreneurial talent and sustain business (ILO, 2002). When young entrepreneurs were asked to describe how they had acquired skills related to their business, almost all reported having done so through informal channels such as exposure during employment or by working closely for or with relatives and other people. Some young entrepreneurs had started by engaging in the business run by a close relative or friend, rather than following their own interests. In such cases, they had gained experience before starting their own business. Unfortunately, in some instances, their employers had shown reluctance to let them stand on their own feet. Mutale's case illustrates this. He had learnt many of his skills helping his brother (a mechanic) whose business was based at their parents' home. When Mutale told his brother he had decided to start his own business, the latter was, unfortunately, very reluctant to help him:

> I shared my intentions to start a business of my own with my brother. But from his talk, and also his behaviour later on, it was clear that my brother did not want me to be on my own in business. He wanted me to continue working with him yet he was not giving me enough money whenever I asked for help from him. Meanwhile we were making a lot of money and sometimes it was me who used to handle the money we made from business.

Jackson had learnt shoemaking by spending time with an old man who used to repair shoes near his home:

> My primary reason of spending time with this old man had nothing to do with learning skills of how to make shoes. It was simply a matter of passing time out of boredom. My dream then was to become a truck driver. Even when I came to live in Lusaka that was what I thought I would do. The old man used to involve me to perform simple tasks, out of his own convenience. Little did I realise that I was learning the art and acquiring skills that would later become a basis for my livelihood income.

For some entrepreneurs such as Tobias, even engaging in casual conversation with friends had helped him learn survival skills that he now employed in his business. In Anna's case, it was partly her own mother and her employer (her aunt's friend) who had taught her the business skills she now used: "I knew nothing about the restaurant business, besides cooking basic traditional meals, which I learnt from my mother at home."

In terms of skills, we see a combination of strong ties (friends and family) and weak ties being used as important channels through which young entrepreneurs had acquired their business skills.

Access to customers

Young entrepreneurs said they considered access to customers very important for succeeding in business in the environment in which they operated. Almost all the 10 entrepreneurs to whom I spoke said they had a loyal clientele who supported their business.

> In the market like this one where almost everyone is selling something, you cannot manage if you do not have personalised customers.
>
> (Chiti)

> I have people who come directly to buy from me ... sometimes my customers send children to me saying: go to him and buy such a product; if he does not have it, he will know what to do, meaning if I do not have the item [my] customer wants, I have to look for it on their behalf among other marketers.
>
> (Tobias)

> In this business, you have to be good. If you do not treat customers well, you will not only lose those particular ones, but also their friends. But again, if you are good, you will see customers referring their friends to you.
>
> (Mary)

Young entrepreneurs use different ways to maintain and strengthen network ties outside their friend and family circles. These approaches centre on the notion of "being good", which can mean, among other things, selling at below normal prices, providing *mbasela* (a Nyanja word for "extra"), giving tips, extending credit to known and trusted customers, looking for products elsewhere when one does not have what a customer wants, negotiating prices on behalf of customers when buying from others, and in some instances delivering products to customers.

> When you impress customers, they appreciate. They even start referring other people to you and you will have a lot of customers. And if you do not assist your customers, you lose out in the process. When customers run away from you this way, you have no one to blame but yourself.
>
> (Tobias)

While "being good" seems to serve as insurance against business failure among young entrepreneurs, there are also instances where reciprocity can pose a problem, with some actors making their own choices based on what suits them. For instance, Mary explained how one of her regular customers, an elderly woman to whom she had once extended credit, started ignoring her calls. She would answer when Mary called from a different number but cut the call once she realised it was Mary. Innocent narrated similar experiences with customers who had taken to avoiding him after he had extended favours to them. To some extent, the marginalised position of youth puts them at a disadvantage, as Innocent observed: "Some people are just manipulative and only interested in serving their own interests and take advantage of my age." In Zambian culture, respect for age is a norm that puts young people beneath everyone else in society.

Conclusion

This study has demonstrated the use of social capital in mobilising various resources among young business entrepreneurs in Lusaka's informal economy. Granovetter's social embeddedness is reflected in personal relationships within the radius of individual young entrepreneurs in their given social context, playing an important role in their daily work and accomplishments. Social capital not only provides resources such as information, finance, skills, and customers, but it also helps young entrepreneurs discover opportunities. What emerges in the study is young entrepreneurs' dominant reliance on strong ties; however, it is also interesting to note that, even with strong ties such as the family, some young entrepreneurs find it difficult to obtain support for their business. This points to the limits of social capital as a resource for entrepreneurship promotion.

We have also seen the need for young entrepreneurs to maintain and strengthen network ties by cultivating connections, particularly those outside their family circles, by "being good". This implies going beyond their business in extending favours. Considering that social capital involves two parties – in this context young entrepreneurs themselves and the people to whom they are connected – either actor might choose to look at the same connection from her or his own standpoint in terms of potential benefits. While such connections are based on some kind of trust, reciprocity should not always be taken for granted. Miscalculations can arise, leading to increased transaction costs to the disadvantage of the young entrepreneur.

Social capital can indeed make it possible for people to achieve certain ends that might not otherwise be attainable (Burt, 2001: 32), but it can also act as a barrier to entrepreneurial development. Young entrepreneurs, therefore, cannot rely solely on social capital as a basis of support. There is a need to cease considering entrepreneurship in isolation and, when designing policies, pay greater attention to understanding the contextual issues surrounding young people in their own environment. In this regard, official support from the government is crucial if entrepreneurship is to make an effective contribution to reducing unemployment and vulnerability among young Zambians.

References

Amorós, J. E., & Bosma, N. (2014). *Global Entrepreneurship Monitor: 2013 global report*. London: Global Entrepreneurship Research Association.

Burt, R. S. (2001). Structural holes versus network closure as social capital. In N. Lin, K. S. Cook, & R. S. Burt (Eds), *Social capital: Theory and research* (Sociology and

Economics: Controversy and Integration series, pp. 31–56). New York, NY: Aldine de Gruyter.

Chigunta, F. (2007). *Investigation into youth livelihoods and entrepreneurship in the urban informal sector in Zambia.* Unpublished PhD thesis, University of Oxford.

Deakins, D., & Freel, M. (2009). *Entrepreneurship and small firms* (5th ed.). London: McGraw-Hill.

Granovetter, M. (1973). The strength of weak ties. *American Journal of Sociology, 78*(6), 1360–1380.

Granovetter, M. (1985). Economic action and social structure: The problem of embeddedness. *American Journal of Sociology, 91*(3), 481–510.

Granovetter, M. (1992). Economic institutions as social constructions: A framework for analysis. *Acta Sociologica, 35,* 3–11.

Hansen, K. T. (2010). Changing youth dynamics in Lusaka's informal economy in the context of economic liberalisation. *African Studies Quarterly, 11*(2–3), 13–27.

International Labour Office (ILO). (2002). *International Labour Conference, 90th session, Report IV: Decent work and the informal economy.* Geneva: ILO.

Locke, C., & Verschoor, A. (2008). *The economic empowerment of young people in Zambia* (Report No. 51431). Washington, DC: World Bank.

Mulenga, L. C. (2000). *Livelihoods of young people in Zambia's Copperbelt and local responses.* Unpublished PhD thesis, University of Wales, Cardiff.

Mwenechanya, S. K. (2007). *Legal empowerment of the poor: Empowering informal businesses in Zambia* (Issue paper prepared for United Nations Development Programme [UNDP] Commission on Legal Empowerment of the Poor). Lusaka, Zambia: UNDP.

Narayan, D., & Cassidy, M. F. (2001). A dimensional approach to measuring social capital: Development and validation of social capital. *Current Sociology, 49*(2), 59–102.

Nchito, W. (2006, September). *A city of divided shopping: An analysis of the location of markets in Lusaka, Zambia.* Paper presented at the 42nd ISOCARP Annual World Congress, Istanbul, Turkey.

Roever, S. C. (2005). *Negotiating formality: Informal sector, market, and the state in Peru.* Unpublished PhD thesis, University of California, Berkeley.

Tanga, P. T. (2009). *Informal sector and poverty: The case of street vendors in Lesotho.* Addis Ababa: Organisation for Social Sciences Research in Eastern and Southern Africa.

van der Leij, M. J. (2006). *The economics of networks: Theory and empirics.* Unpublished PhD thesis, Erasmus Universiteit, Rotterdam.

Woolcock, M., & Narayan, D. (2000). Social capital: Implications for development theory, research, and policy. *World Bank Research Observer, 15*(2), 225–249.

16 Innovative approaches by Ugandan microfinance institutions to reach out to young entrepreneurs

Agnes Noelin Nassuna, Søren Jeppesen, and Waswa Balunywa

Introduction

Youth unemployment is one of the major challenges facing economies in sub-Saharan Africa (International Labour Organization, 2012). A key intervention in this context is youth entrepreneurship, where young people are encouraged to start their own business (Awogbenle & Iwuamadi, 2010). Entrepreneurship efforts, however, are unlikely to succeed without easy access to finance. Although commercial banks serve as many entrepreneurs' main source of finance, young people tend to lack the collateral – assets such as land, automobiles, real estate, or investments – needed to access this source. This constraint to finance exacerbates the youth unemployment problem (World Bank, 2009). In this context, microfinance institutions (MFIs) have emerged as an important innovation in the financial sector, offering financial services to the unbanked and the vulnerable (Ledgerwood, 2000). Among vulnerable groups, MFIs have tended to focus on women, who are assumed to be trustworthy and represent a higher repayment probability (Brau & Woller, 2004; Kato & Kratzer, 2013).

MFIs rarely extend services to young people, whom they associate with a high risk of loan non-repayment (Donahue, James-Wilson, & Stark, 2006). MFIs also tend to presume that young people are not creditworthy, given their limited business experience and knowledge and their low levels of social and financial power (Clemensson & Christensen, 2010). Shah, Mohamed, Saraf, and Nigam (2010) find that 60% of MFIs in Africa do not lend to youth; the remaining 40% claim not to discriminate by age but, in practice, do not serve young people. In spite of the generally low interest in serving youth, some few MFIs have gone on to provide specific services to young people. The purpose of this chapter is to examine the innovative financing approaches that some MFIs in Uganda have adopted in order to serve youth through financial and social intermediation.

MFIs and the need for youth tailored innovations

MFIs are primarily financial intermediaries for low-income and vulnerable groups who cannot obtain financial services from formal commercial banks (Hermes & Lensink, 2007; Ledgerwood, 2000). Without an intermediary, such people would be exposed to information asymmetries, bounded rationality, and opportunistic behaviour that characterise an imperfect market (Allen & Santomero, 1998). Information asymmetries arise when people fail to engage successfully in financial transactions because they lack adequate information about the intentions and ability of others to honour their obligations (Clarkson, Jacobsen, & Batcheller, 2007). In an imperfect market, some economic actors may take advantage of others out of self-interest (opportunistic behaviour) while other economic actors make irrational decisions due to limited information, time, and cognitive ability (bounded rationality). These challenges make access to finance cumbersome and costly, affecting youth more so because they lack the financial skills, collateral, and business knowledge to bolster their economic position (Garcia & Fares, 2008). This increases the relevance of and need for intermediaries with the capacity to minimise such challenges (Spulber, 1996), especially for young people. If youth cannot access services from commercial banks, their only alternative remains the MFIs.

Microfinance services were originally offered by nongovernment organisations (NGOs), which focused on extending credit to the poor through microloans for business (Ledgerwood, 2000). In addition to financial intermediation, MFIs were set up to act as social intermediaries that enabled vulnerable people to gain value from the money they had borrowed (Kistruck, Beamish, Qureshi, & Sutter, 2013). Social intermediation involves activities aimed at building self-confidence, social capital, financial literacy, and management skills among borrowers and savers (Ledgerwood, 2000). Many MFIs have prioritised business loans over other financial products needed by the poor. With time, other institutions such as banks, private companies, micro-deposit-taking institutions (MDIs), and savings and credit cooperative organisations (SACCOs) have emerged to offer microfinance services ranging from savings, non-business loans, and insurance to transfer payments (Brau & Woller, 2004). Clearly, the poor need not only microloans, but also other services offered by the financial sector. The number of players has increased competition within the microfinance sector globally and the corresponding use of technological advancements such as automatic teller machines, mobile banking, and point-of-sale machines has also escalated.

Overall, MFIs have achieved a number of key goals in terms of high repayment rates and outreach to women (Sengupta & Aubuchon, 2008) – a global survey of over 1,000 MFIs shows that 75% of their clients are women (Adams, Awimbo, Goldberg, & Sanchez, 2000). Despite this, MFIs still face various challenges that threaten their performance and weaken their drive to reach out to the poor. Many MFIs have drifted away from their original mission of serving the poor (Hermes, Lensink, & Meesters, 2011) with a resulting trade-off between social goals and financial gains. Their outreach remains limited, especially where vulnerable groups such as youth are concerned (Donahue et al., 2006). Additionally, many MFIs operate at very high costs, making them inefficient enterprises (Sengupta & Aubuchon, 2008).

In order to surmount their performance challenges and function as effective financial and social intermediaries, especially for the youth population, MFIs need unique capabilities and resources. The first of these is innovation: the intentional introduction and application of ideas, processes, products, or procedures new to the relevant unit of adoption (Johannessen, Olsen, & Lumpkin, 2001). An innovation need not be an invention; as long as it is new within the confines of that unit, it is an innovation. Innovations, in turn, need an intrapreneurial environment if they are to succeed. An intrapreneurial environment is the sum of internal organisational factors that act as antecedents to entrepreneurship within an organisation (Hornsby, Kuratko, & Zahra, 2002). These include the organisation's disposition and capacity to innovate, i.e. its management practices and resources geared towards innovation (Amabile, 2013).

In Uganda, MFIs developed as a response to the collapse of the financial sector in the 1980s. By the 1990s, the financial sector had begun to recover while MFIs flourished alongside. By the 2000s, the microfinance sector had become a prominent part of the financial sector (Goodwin-Groen, Bruett, & Latortue, 2004). Part of their success is explained by the high percentage of the adult population – over 79% according to Synovate (2010) – that cannot access financial services from formal commercial banks. This large unbanked population thus provides potential clients for MFIs.

MFIs in Uganda range from banks and MDIs to SACCOs, NGOs, and companies. Bank of Uganda categorises MFIs under four tiers: tier one comprises banks that offer microfinance services, tier two consists of credit institutions, tier three includes MDIs, and tier four comprises all others, such as SACCOS, NGOs, and smaller companies (Association of Microfinance Institutions in Uganda [AMFIU], 2008a). By 2009, the microfinance sector comprised two banks, three MDIs, and 70 registered SACCOs; the remaining institutions were unregulated and unregistered (Kumwesiga, 2009). Regulated and registered Ugandan MFIs come together under the AMFIU, an apex body that pursues best practices in microfinance. It is worth noting that, although unregulated and predominantly unstructured, the MFIs in tier four serve the majority of Ugandans.

MFIs have proven tremendously successful in Uganda, especially in terms of outreach (AMFIU, 2008b). The regulated MFIs reach about 4% of the population while those that are unregulated reach about 50% (Synovate, 2010). Similar to the global picture, studies show that MFIs in Uganda have also helped empower women, who account for over 70% of their clients (Kumwesiga, 2009). The main product initially offered by Ugandan MFIs was microcredit, but over time, they have introduced other financial products including savings, insurance, and money transfer. However, as mentioned earlier, many of these MFIs operate at very high costs, which threaten their sustainability (Kamukama, Ahiauzu, & Ntayi, 2010).

MFIs in Uganda also tend to overemphasise financial performance – charging very high interest rates of over 30% – and neglect their social performance.

Additionally, they have not tapped the youth market, despite the country's large percentage of unemployed youth. MFIs could play a very important role in mitigating Uganda's unemployment problem if they were to provide unemployed youth with affordable finance that would enable them to venture successfully into entrepreneurial activities.

Methodology

This chapter is based on cases investigated as part of a wider study on the intrapreneurial environment and performance of MFIs. The wider study used a mixed-methods approach involving both quantitative and qualitative designs. The qualitative design focused on investigating cases that would generate a deeper understanding of how the intrapreneurial environment and innovation influenced performance. These case studies were drawn from a survey of 113 MFIs in Uganda, of which 10 were selected because they had employed a unique innovation that was perceived to have greatly influenced their performance. The names of the participating organisations and products were changed to ensure confidentiality. Out of the 10 case studies, three were providing services that could be termed unique for a youth clientele. It is these three cases (referred to here by pseudonyms) that form the gist of this chapter, which investigates how the innovations were initiated and which internal environmental factors influenced their success. In addition, the intermediation taking place and the impact of the innovation on performance are examined.

Three different managers from each of the organisations selected were interviewed by the first author, including the overall chief executive or chairperson of the board, the manager in charge of the innovation selected, and the initiator of the innovation, where available. In order to verify the information gathered at the organisation level, focus group discussions were held with customers to examine the general benefits and challenges of the innovation concerned. Each discussion comprised five customers (from the respective organisation) who were directly using the identified innovation. The discussions were held at the customers' regular meeting sites.

The data collected was then analysed using the software Nvivo 10. This involved extracting the data and analysing the relevant themes in order to contextualise the information provided by the interviews in relation to the study's major constructs, i.e. the intrapreneurial environment, innovation, intermediation, and performance. Our field notes helped capture the initial interpretation of the findings during data collection. The field notes were also entered into Nvivo and interpreted alongside the data from the main interviews.

Case studies of innovative microfinance institutions in Uganda

The following sections provide a snapshot of the innovations investigated and describe how each was initiated, developed, and implemented within each of the three organisations.

ABC Ltd: a youth savings initiative

ABC Ltd is a Ugandan MDI that offers microloans to low-income earners. It started as a women-only MFI but later opened up to men. In 2009, the organisation introduced a savings product known as "Shine up" for young girls between the ages of 13 and 24 as they were assumed to be more vulnerable. In 2011, the product was also made available to boys. "Shine up" encourages youth to save by training them in financial literacy and showing them how to open a savings account. The product was implemented in schools and local communities where teachers, community leaders, and elders acted as the young people's mentors. Within the community, ABC organised community days in conjunction with local leaders in order to meet young people and their parents. When targeting the schools, ABC sought the head teachers' permission to introduce the students and their teachers to the product. "Shine up" clients would then meet every week to give their mentors the money that was to be deposited in the bank and to receive some financial training. Additionally, inter-club games, competitions, and parties were held for the young people involved.

To ensure the success of the product, ABC management sensitised their existing staff and recruited younger staff to train their new clientele group. A field officer who had shown an extraordinary interest in "Shine up" was asked to spearhead ABC's efforts to implement, and subsequently manage, the product. The organisation also sought external support from ABC International, its parent organisation, to further ensure the successful implementation of the product. However, in the process, the MFI's funding from ABC International was reduced and it began to finance the product locally. This led to a reduction in some of the activities it had initiated, such as annual parties, inter-club games and competitions for youth clients.

Clients report that the product has had a number of benefits: they have learnt to accumulate savings over time and to use these to top up school fees or start a business; they have also acquired financial discipline that enables them to forgo impulsive purchases (and save instead) and to develop clear business and financial goals. Young people also report facing several challenges: as mentioned above, ABC's youth activities have been scaled down while the non-availability of mentors and the failure of young people's peers to join the clubs (due to the lack of guardians or parents) means that the weekly depositing facility has become less regular. Compulsory weekly savings can also be difficult sometimes if young people have no money to deposit. Despite these challenges, ABC reports that the product is deemed successful and that it is committed to developing "Shine up" in all its branches in the future. It also hopes to increase the flexibility of the product, for example, by removing the weekly savings deposit rule, especially for rural youth.

OPQ Uganda: a holistic approach to serving youth

OPQ Uganda is an NGO-based MFI with over 90 branches across the country. It has initiated a holistic product known as "Empowerment and Livelihood for

Adolescents" (ELA), which targets young people between the ages of 13 and 22. The product involves six components: games, music, and drama; life and health skills; livelihood and job creation; financial literacy; vocational training; and microfinance. On completion of all six components, youth are encouraged to start their own business while some are employed in certain positions with OPQ's partner organisations.

The product was adapted from OPQ's headquarters in Brazil, which provided the initial funding. Originally aimed at girls, ELA was made available to boys in 2012. The product focuses on young school dropouts in local communities and is implemented through the formation of youth clubs. OPQ approaches potential clients through local council leaders who mobilise the area's youth. From there, young people are encouraged to form clubs and select their own leaders. An OPQ field assistant officer is attached to each club and provides mentorship and guidance. The organisation has also created two new internal positions at its head office – a director ELA and a job creation manager, both of whom are responsible for ensuring the successful implementation of the product.

OPQ charges interest on the loans it gives to the youth under the microfinance component; the other five components of the product may be utilised free of charge. ELA clients report having benefited from the product in different ways. The life-skills training and sex education, for instance, has encouraged them to avoid casual sex and unwanted pregnancies. They claim to have acquired practical skills (in tailoring, hairdressing, carpentry, computer use, and poultry rearing) and financial skills (budgeting, saving, investing, and credit management). Clients also indicate they now have easier access to finance and are allowed to borrow without collateral. Nonetheless, young people report that the loan amounts are small and that they cannot remain members of the youth clubs after the age of 24. Funding from OPQ's headquarters has also ceased, affecting some of the organisation's activities. Nonetheless, it is committed to seeing the programme succeed.

XYZ: providing decent housing for youth

XYZ is a rural SACCO based in one of the most remote districts of Uganda. In 2006, its chairperson, himself a young adult, initiated a youth start-up loan product intended to enable young people to restart their lives and acquire acceptable housing facilities. The chairperson had found that many young people who had decided to leave home were now living in ramshackle structures; some were idle and engaged in drug abuse, theft of farm produce, and casual sex. Originally, the product focused on the provision of physical household items and construction materials on a loan basis. Clients were required to pay back the cost of the items received in instalments at a small interest fee of 6%. The product has now evolved to include a savings component, mentorship, and job creation.

While implementing the product, XYZ set up a committee to identify, mobilise, and organise youth training sessions. The committee approached young people through village meetings at which they explained their product to village members. They were also tasked with the responsibility of purchasing the household items and identifying on-the-job mentors. Over time, XYZ found it had insufficient funds to extend to its clients and, as a result, the SACCO's adult members agreed to contribute an additional amount to create a special fund for youth.

The organisation's clients reported that the product has had a number of benefits. Many young people said they have been able to reconstruct their houses and get married since they were no longer ashamed of their homes. Others said they now have improved access to business assets (such as oxen or motorcycles) on a hire-purchase basis, which has helped them expand their businesses and engage in commercial agriculture. In turn, the SACCO's members provide a readymade market for their produce. The key challenges that clients face include the impact of inclement weather on farming, which can make it difficult to repay loans if a client is unable to produce and sell enough; and situations in which some clients may choose to default on their loans and abscond, leaving their peers responsible for returning the money. Nonetheless, XYZ is hopeful that the programme will continue.

Role of innovation in facilitating financial and social intermediation for young people

Drawing on these three case studies, this section examines the nature of each innovation – the newness of the product or service and how it has facilitated financial and social intermediation – and the underlying intrapreneurial environment, which, in turn, affects product implementation and organisational performance.

All three products – "Shine up", ELA, and the start-up loan for housing and asset purchase – were new in the context of that particular organisation and its clients, and had been introduced in response to a community need or crisis. Johannessen et al. (2001) address the issue of innovation newness by explaining that "how new" depends on "new to whom", that is, the unit or organisation introducing or adopting the innovation. The products in these cases were specifically introduced for young people. This is contrary to the observations of James-Wilson and Hall (2006), who have found that MFIs neither actively target youth nor see them as a separate market.

ABC's savings product is clearly different from its mainstream loan products for older clients. The product arose as a result of research that showed that many young people needed assistance but lacked the influence of a culture of saving and were unable to use their money efficiently. OPQ's product was a response to the high dropout rates among youth. ELA is unique because it provides not only financing, but also life and health skills, a chance to discover inherent talent, vocational skills, and job creation – all of which are aimed at making youth better clients for the future. XYZ's product was a response to the plight of young people who had sought independence but had nowhere to live. Instead of adopting a conventional lending approach, XYZ provides physical items that enable young people to construct habitable structures and then mentors them in starting their own businesses. Although James-Wilson and Hall (2006) note that MFIs avoid financing start-up businesses, OPQ and XYZ do just this.

In all three cases, there is some form of financial intermediation focusing on youth. All three products help young people access financial services at different levels, thereby reducing the financial challenges they might otherwise face (as reported by Donahue et al., 2006; Awogbenle & Iwuamadi, 2010). ABC helps youth mobilise their savings at the school and community level on a non-profit basis. Although the traditional role of MFIs is to offer credit (Ledgerwood, 2000), savings mobilisation is key in the three case studies. The underlying hope is that their accumulated savings will enable youth to acquire wealth and use it to access other assets. In addition, these savings add to the deposits of the organisations and can then be used to extend credit. Although these deposits are small, they are expected to build up over time. Moreover, OPQ and XYZ offer credit after they have provided young people with some form of training and encouraged them to save. This holistic approach to financial intermediation thus provides youth with a variety of financial services.

All three cases indicate efforts to build the capacity of young people to engage meaningfully and actively in financial services. This is the crux of social intermediation (Edgcomb & Barton, 1998; Kistruck et al., 2013; Ledgerwood, 2000) and was carried out mainly through training. ABC trains young people in financial literacy and purposive saving. OPQ trains them in various spheres beyond the financial, including health, games, music, and drama; it has also sent some youth to vocational schools. The rationale is that this holistic approach will equip young people with better skills to deal with the credit they finally acquire. XYZ uses mentorship as a form of training whereby adult members employ youth in their respective businesses for one year. The programmes also help young people engage productively in some form of economic activity. The case studies show that youth who work together in groups tend to enhance their social capital, which yields various benefits in terms of exposure, self-confidence, and information sharing. The organisations all hope that such initiatives will make young people better entrepreneurs.

The key intrapreneurial environmental factors observed in these cases include management support, the ability to raise funds for the product, training, and mentorship. Management support seems to be a critical factor across all three cases and includes initiating research, identifying a champion, creating an innovation team, and mobilising other managers, staff, and clients. At ABC, the idea of supporting youth had been introduced earlier, but the product did not materialise until the management sanctioned research on the innovation, identified someone to spearhead the project, and recruited staff to implement the product. In the case of OPQ, the idea was adopted from abroad (the organisation's headquarters), but was successful only when the managers in Brazil sought the support of their colleagues in Uganda. On the other hand, XYZ's product was initiated by the chairperson himself who later set up a management committee to spearhead the implementation of the product. Another key management effort observed was the need to actively seek staff and clients' support. Mobilising staff and clients critical to ensure the success of the innovation and failure to do so may have hindered the product's chances of succeeding. We saw this in the case of OPQ, which had failed to mobilise clients in the remote area of Karamoja.

The ability to acquire funding specifically for the innovation concerned was another key intrapreneurial factor in all three cases. ABC and OPQ had external funding available, which enabled them to implement the innovation, but both were compelled to reduce the scope of their youth activities as this source of funding dried up. Subsequently, OPQ tried to encourage youth to fund their activities and take ownership of the product in a bid to make it self-sustaining. In XYZ's case, no external funding was available to support the innovation and its members had to mobilise their own funds. Training – whether for employees, clients, or members – was another important internal factor that determined the successful initiation and implementation of these innovations. XYZ's training was supplemented with a mentorship scheme whereby youth underwent experiential learning facilitated by the organisation's adult members. This helped them start their own businesses after a year of mentorship.

The case studies show that each innovation has influenced the performance of its organisation. In all three cases, the organisations have increased their customer base as a result of the new products. ABC, for instance, has created a new niche of young customers influenced by the product's culture of saving and it foresees them becoming better clients than its existing adult clientele. By intermediating for young people, ABC has also gained other clients – their parents and guardians who not only save with them but also borrow money. The products have also earned a small profit for their organisations. OPQ and XYZ have benefited directly from the services they extend: both charge an interest on youth loans, thus earning some profit. ABC has not benefited directly in financial terms from the product, but it does expect to earn profits in the future as its youth clients mature into responsible adult clients.

Conclusion

The three case studies above show that financial intermediation for youth is possible and not necessarily risky. It is made possible by new and unique financial products that are tailored to young people's needs. As Johannessen et al. (2001) explain, innovation can be used as a strategy to enter unexploited areas – in this case, focusing on youth as unique clients.

The importance of social intermediation emerges clearly in the cases presented and implies that it should be embedded in the financial products to enable youth to engage more meaningfully in entrepreneurship. The skills they gain through social intermediation – self-confidence, financial literacy, and business management – also increase their ability to escape unemployment. This emphasis is similar to the argument made by Kistruck et al. (2013), who explain that social intermediation builds the capability of participants in the intermediation process to engage more meaningfully in financial transactions. This chapter has shown that training and mentorship are practical ways of helping youth acquire financial skills that they can then use to start and effectively manage a business. MFIs could, therefore, embed training and mentorship into their financial products for youth. Moreover, this could be done in partnership with successful businesses and existing entrepreneurs.

The chapter has also shown that internal environmental factors influence the successful implementation of innovation. This is consistent with other studies that show how such an environment stimulates innovation (see Amabile, 2013; Hornsby et al., 2002). The case studies emphasise the importance of management support, training, finances, and staff and client mobilisation as key internal factors that enable MFIs to implement successful innovations. Moreover, MFIs are capable of serving youth in a sustainable manner if they focus on improving their intrapreneurial environment and innovations. Governments and donors, for instance, could inject funds into youth-centred MFI programmes to help create special funds for promoting such innovations and facilitate training.

Finally, youth-tailored financial products can also be profitable, as in the case of OPQ and XYZ. This observation is similar to that of earlier studies carried out by Donahue et al. (2006) as part of USAID's efforts to investigate how MFIs might reach out to young people. Given that many other MFIs are reluctant to handle a youth clientele, awareness programmes and training for their management and staff could prove very useful.

References

Adams, L., Awimbo, A., Goldberg, N., & Sanchez, C. (Eds). (2000). *Empowering women with microcredit* (Microcredit Summit Campaign Report 2000). Washington, DC: Microcredit Summit Campaign.

Allen, F., & Santomero, A. M. (1998). The theory of financial intermediation. *Journal of Banking and Finance, 21*, 1461–1485.

Amabile, T. (2013). Componential theory of creativity. In E. Kessler (Ed.), *Encyclopaedia of management theory* (pp. 135–140). Thousand Oaks, CA: Sage.

Association of Microfinance Institutions of Uganda. (2008a). *Uganda microfinance industry assessment.* Kampala: AMFIU.

Association of Microfinance Institutions of Uganda. (2008b). *Microfinance tomorrow: Refocusing the vision for the industry in Uganda.* Kampala: AMFIU.

Awogbenle, A. C., & Iwuamadi, K. C. (2010). Youth unemployment: Entrepreneurship development programme as an intervention mechanism. *African Journal of Business Management, 4*(6), 831–835.

Brau, J. C., & Woller, G. M. (2004). Microfinance: A comprehensive review of the existing literature. *Journal of Entrepreneurial Finance and Business Ventures, 9*(1), 1–26.

Clarkson, G., Jacobsen, T., & Batcheller, A. (2007). Information asymmetry and information sharing. *Government Information Quarterly, 24*(4), 827–839.

Clemensson, M., & Christensen, J. D. (2010). *How to build an enabling environment for youth entrepreneurship and sustainable enterprises* (Paper for the knowledge sharing event on Integrated Youth Employment Strategies, February, Moscow). Geneva: International Labour Office.

Donahue, J., James-Wilson, D., & Stark, E. (2006). *Microfinance, youth and conflict: Central Uganda case study* (microREPORT No. 38). Washington, DC: United States Agency for International Development.

Edgcomb, E., & Barton, L. (1998). *Social intermediation and microfinance programs: A literature review*. Bethesda, MD: Microenterprise Best Practices Project.

Garcia, M., & Fares, J. (Eds). (2008). *Youth in Africa's labour market*. Washington, DC: World Bank.

Goodwin-Groen, R., Bruett, T., & Latortue, A. (2004). *Uganda microfinance sector effectiveness review*. Washington, DC: Consultative Group to Assist the Poor.

Hermes, N., & Lensink, R. (2007). The empirics of microfinance: What do we know? *The Economic Journal, 117*(517), 1–10.

Hermes, N., Lensink, R., & Meesters, A. (2011). Outreach and efficiency of microfinance institutions. *World Development, 39*(6), 938–948.

Hornsby, J. S., Kuratko, D. F., & Zahra, S. A. (2002). Middle managers' perception of the internal environment for corporate entrepreneurship: Assessing a measurement scale. *Journal of Business Venturing, 17*(3), 253–273.

International Labour Organization. (2012). *Africa's response to the youth unemployment crisis: Regional report*. Geneva: International Labour Office.

James-Wilson, D., & Hall, J. (2006). *Microfinance, youth and conflict: West Bank case study* (microREPORT No. 41). Washington, DC: United States Agency for International Development.

Johannessen, J.A., Olsen, B., & Lumpkin, G. T. (2001). Innovation as newness: What is new, how new, and new to whom? *European Journal of Innovation Management, 4*(1), 20–31.

Kamukama, N., Ahiauzu, A., & Ntayi, J. M. (2010). Intellectual capital and financial performance in Ugandan microfinance institutions. *African Journal of Accounting, Economics, Finance and Banking Research, 6*(6), 17–32.

Kato, M. P., & Kratzer, J. (2013). Empowering women through microfinance: Evidence from Tanzania. *ACRN Journal of Entrepreneurship Perspectives, 2*(1), 31–59.

Kistruck, G. M., Beamish, P. W., Qureshi, I., & Sutter, C. J. (2013). Social intermediation in base-of-the-pyramid markets. *Journal of Management Studies, 50*(1), 31–66.

Kumwesiga, C. T. (2009). *State of microfinance in Uganda 2008: Analysing AMFIU member MFIs* (Working Paper No. 10). Kampala: Association of Microfinance Institutions of Uganda.

Ledgerwood, J. (2000). *Microfinance handbook: An institutional and financial perspective*. Washington, DC: World Bank.

Sengupta, R., & Aubuchon, C. P. (2008). The microfinance revolution: An overview. *Federal Reserve Bank of St. Louis Review, 90*(1), 9–30.

Shah, A., Mohammed, R., Saraf, N., & Nigam, R. (2010). *Microfinance services for youth in the sub-Saharan African region* (Mimeo). Coventry: Warwick Social Enterprise and Microfinance Society.

Spulber, D. F. (1996). Market microstructure and intermediation. *Journal of Economic Perspectives, 10*(3), 135–152.

Synovate. (2010). *Results of a national survey on demand, usage and access to financial services in Uganda*. Kampala: Finscope.

World Bank. (2009). *Africa development indicators 2008/2009*. Washington, DC: World Bank.

17 Entrepreneurship education in Uganda

Impact on graduate intentions to set up a business

Edith Mwebaza Basalirwa, Katherine V. Gough, and Waswa Balunywa

Introduction

Education plays an important role in the development process, changing people's attitudes and giving them knowledge and skills that enable them to change their lives. Although education levels in the global South are improving (UNESCO Institute for Statistics, 2011), many young people find it very difficult to obtain employment (International Labour Organization, 2009). Governments, donor organisations, and nongovernment organisations have introduced various programmes to tackle youth unemployment including entrepreneurship education (De Jaeghere & Baxter, 2014). Entrepreneurship education is seen as an avenue for job creation as it should provide students with the requisite skills and stimulate their intention to start a business. Given the perception of the importance of entrepreneurship, at both the macro level of economic development and at the micro level of personal satisfaction and achievement (Alberti, Sciascia, & Poli, 2004), and the recognition that entrepreneurial ability can be increased through education (Gorman, Hanlon, & King, 1997; Ronstadt, 1987), there is rising interest in developing educational programmes to encourage and foster entrepreneurship. Although entrepreneurship education is a relatively new field, there is growing recognition that it can contribute towards the creation of an enterprise culture among students (Kuratko, 2003; Solomon, Duffy, & Tarabishy, 2002).

While the effect of education in general on employment opportunities has been explored in a sub-Saharan African context (UNESCO Institute for Statistics, 2011), there are few studies on the effect of entrepreneurship studies particularly at university level (however, see Byabashaija & Katono, 2011). Most studies examining students' attitudes towards entrepreneurship and the barriers they face focus on countries in the global North (Gnoth, 2006; Kolvereid, 1996) or in the newly rising Asian states (Sandhu, Sidique, & Riaz, 2011; Wang & Wong, 2004). This chapter aims to contribute to debates concerning university entrepreneurship education in sub-Saharan Africa by examining who studies entrepreneurship education and why, their career paths after graduating, and how their studies have impacted on their entrepreneurial intentions. It draws on a study of recent graduates from two universities in Uganda who had studied entrepreneurship.

Entrepreneurship education

Entrepreneurship education lacks a common definition but is generally considered to be "the structured formal conveyance of entrepreneurial competencies, which in turn refer to the concepts, skills and mental awareness used by individuals during the process of starting and developing their growth-oriented ventures" (Alberti et al., 2004: 5). Although research into entrepreneurship education is limited, existing studies indicate that exposure to certain types of entrepreneurship education, especially the practical parts of programmes, affect attitudes to entrepreneurship and enhance an individual's intention to start a business (Honig, 2004). Since entrepreneurship can positively affect economic growth and development, it is argued that governments should attempt to increase the supply of entrepreneurs including by initiating entrepreneurship educational programmes (Thurik & Wennekers, 2004).

Many business schools have introduced entrepreneurship education into their programmes in response to this perception of the importance of entrepreneurship as a solution to the graduate youth unemployment problem. Key to understanding the impact of entrepreneurship education at university level is an analysis of the intentions of graduates to set up a business. This study draws on Ajzen (1991), who argues that attitudes, subjective norms, and perceived behaviour control are predictors of behavioural intention. "Attitude" is what an individual thinks about a certain behaviour; it is a personal evaluation, a result of appraising behaviour and deciding whether it is good or not. "Subjective norm" is the socially expected norm of conduct; it refers to the degree to which family, friends, peers, and society at large expect or pressurise an individual to perform the behaviour in question. If the environment in which a person is embedded is highly supportive of entrepreneurial activity, he or she will more likely pursue this as a career option (Byabashaija & Katono, 2011). Perceived behavioural control refers to the extent to which an individual feels capable of a specific behaviour; it is based on the person's know-how and experience and their appraisal of any likely obstacles to the behaviour (Ajzen, 1991). The greater the feeling of behavioural control, the stronger the intention to behave in that particular way will be (Byabashaija & Katono, 2011). Graduates from entrepreneurship programmes are shown to be more likely to have entrepreneurial characteristics than those graduating from other subjects, supporting arguments that entrepreneurship education increases the intention of students to start a business (Kolvereid & Moen, 1997).

The Government of Uganda is promoting entrepreneurship education in the hope that this will provide young people with the knowledge and skills to enable them to start businesses. Given that 60% of young people graduating from institutions in Uganda do not find jobs (Uganda Bureau of Statistics [UBOS], 2010),

Figure 17.1 Entrepreneurship teaching at Makerere University Business School (photo: Abdu Nsubuga).

there is an acute awareness of the need to encourage young people to become entrepreneurs. Entrepreneurship is currently taught at secondary school level as well as in colleges and universities, and it is being proposed as a subject at primary school level. While some universities offer courses in entrepreneurship to students of other disciplines, degree-level standalone entrepreneurship education is offered in just two universities: the Makerere University Business School (MUBS) and Uganda Christian University (UCU) (see Figure 17.1). MUBS started teaching an entrepreneurship programme in 2002 with the first students finishing in 2005 and graduating in 2006, while UCU started teaching entrepreneurship in 2009 with the first students completing and graduating in 2011. The average number of students enrolled per year in MUBS is around 165 while at UCU it is about 95. It is the entrepreneurship graduates of these two universities who are the focus of this chapter.

Methodology

This chapter forms part of a wider study on the perceived value of entrepreneurship education on the entrepreneurial proclivity of graduates. The study population consisted of graduates of the Bachelor of Entrepreneurship and Small Business Management at MUBS for the years 2006–2012 and graduates of the Bachelor of Entrepreneurship and Project Planning at UCU for the years 2011 and 2012. Both quantitative and qualitative methods were used to collect data to determine the experiences and opinions of these graduates.

Quantitative data was collected using a self-administered questionnaire during the period April to October 2013. Out of a total of 2,130 graduates from the two entrepreneurship degrees, a sample of 350 (16%) was randomly selected from lists of graduates obtained from the universities. Two research assistants who were former students of MUBS were employed to carry out the data collection. Where possible (in around 100 cases), respondents completed the questionnaire online. Others filled out the questionnaire by hand, either emailing it or handing it back to the research assistants. The questionnaires were divided into four main sections: the characteristics of the respondent (including personal details, university education, and businesses ownership), perceived value of entrepreneurship education, entrepreneurial proclivity, and business start-ups. On average, it took 30 to 40 minutes to complete a questionnaire.

Qualitative data was subsequently collected to gain a deeper insight into who studies entrepreneurship and why, what they do after graduation, and how they perceive their education to have influenced their career paths. Semi-structured interviews were held with a total of 18 graduates who were purposively selected from the questionnaire respondents; two-thirds of the graduates selected were running a business. The interviews were conducted by the first author and lasted between 30 minutes and one hour. They were taped and subsequently transcribed, though their names have been changed to ensure confidentiality. This chapter is developed from the combined quantitative and qualitative data set.

Entrepreneurship graduates

In order to provide a picture of who studies entrepreneurship and why, a profile of the graduates is presented here, including their experience of running a business, followed by an analysis of why they decided to study entrepreneurship and what they have been doing since graduating.

Profile of graduates

The questionnaire survey revealed that slightly more graduates of entrepreneurship are female (56%) than male, which, given that only 40% of graduates overall are female (Kagoda, 2011), indicates that studying entrepreneurship is especially attractive to young women. All the graduates were at least 21 years old and the vast majority (98%) were aged below 36, thus falling within the category of youth, and almost all were Ugandan (95%). Interestingly, the majority of graduates (59%) come from families that run businesses, indicating that many of the young people have grown up in entrepreneurial environments. Juliet (aged 27), for example, indicated how

> I got interest for business via my parents especially daddy, who is self-employed. He is in the businesses of spare parts, motor vehicle garage, and

farming. Mummy is also in business and politics. Now all my family is in business.

Growing up in an entrepreneurial family can result in young people learning business-related skills from an early age. As Rose (aged 25) explained, "I used to go to the village where the family business is to help direct certain things. I would go with our driver to the farm and help with work. This enabled me develop skills at that stage."

Almost half (43%) of the graduates had established a business before even starting their university course, highlighting their entrepreneurial tendencies from an early age. Tom (aged 26), for example, had set up a business selling roast chicken when he was only 15 in order to pay his school fees. He recounted how "I had a business before I started the university entrepreneurship programme and actually the money I got from this business educated me at university. Even when I was still studying at university I had a business running." Similarly, Jane (aged 26) had owned a business making and selling pancakes prior to university; during university, she ran a business selling success cards.[1]

Tom and Jane were not alone in running a business while studying at university; almost two-thirds of the graduates (61%) had run a business concurrently with engaging in their entrepreneurship studies. Many of these were very small businesses that did not require much capital and were easy to run alongside studying. Diana (aged 30) explained how she managed to combine running a business shopping for others with her studies as she only had classes in the morning, which meant she could work in the afternoons. She stopped her business during exam time and then resumed once she had finished. Even small businesses take time to run though, and, as Sarah (aged 25) experienced, this can encroach on study time. As she explained,

> While I was still studying at campus I used to weave pencil bags. The cost of thread and needle was small, the kind you do not need to struggle so much to raise the money. My profit was 70% of the cost. The problem was it would cost me a lot of time, which was very limited then.

This brief profile of students of entrepreneurship shows how many do not wait until after they have graduated to set up a business, but do so before and/or during their studies. Their profits are often used to help finance their studies as most are self-funding students. Hence, on graduating, the majority are not first-time entrepreneurs and come from entrepreneurial families.

Motivation for studying entrepreneurship

Graduates gave a range of reasons for why they had chosen to take a degree in entrepreneurship. The most common reason, given by 45%, was that they considered a degree in entrepreneurship to be the best course available for them (see Table 17.1), either because business was their passion or because they were

Table 17.1 Reasons for choosing an entrepreneurship degree (%)

Reason	Number of students
Most desirable course	45
Employability after the programme	27
Recommended by others	16
Given by admission board	12
Sample size	*350*

Source: Questionnaire survey (2013).

interested in making money. An example is Juliet (aged 27), who indicated how "I chose a course of my own interest. All my life I wanted to be a rich woman so I wanted to be an entrepreneur." Some graduates had already developed an interest in entrepreneurship through studying the subject at secondary school. Noeline (aged 27) explained how she had chosen her course for "the love for entrepreneurship. I started doing entrepreneurship in Senior 1 to Senior 4. I had the passion for doing it.... When I went to 'A' level, I chose it and [then] when going to the university". As Wilson, Kickul, and Marlino (2007) also found, exposing young people to entrepreneurship as an academic subject at an early stage can result in some students being inspired to continue studying it at university.

Just over one-quarter of students (27%) reported that the perceived employability of the programme was their main reason for choosing to study entrepreneurship. Moses (aged 28) explained how "my fear for finding a job ... influenced me so I thought let me be more strategic and have my own business". While some expected employers to be interested in hiring them following graduation, for others it was the desire to be their own boss – often linked to bad experiences of being an employee – that was the driving factor. As Juliet indicated, "I wanted to be my own boss. I first went and worked somewhere and the way they treated us! They mistreated us. So I said one day I should also be a boss." This confirms findings that the desire for independence is an important driver of joining an entrepreneurship course (Morales-Gualdrón, Gutiérrez-Garcia, & Dobón, 2009).

An additional 16% of graduates were influenced by the advice of family, friends, and instrumental figures such as teachers to study entrepreneurship. As Joseph (aged 29) explained, "My class teacher advised me to study entrepreneurship and I did it because it was given to me on government sponsorship." A further 12% indicated that they had not chosen to study entrepreneurship but rather it was allocated to them by the University Joint Admissions Board when they did not succeed in getting their first choice of study. Mathew (aged 26) explained that,

> Actually by the time I decided to fill the form, entrepreneurship was my third choice. First choice was Bachelor of Human Resource Management, second choice was Social Sciences, and Entrepreneurship was the third. I

was admitted onto Entrepreneurship, which I qualified for and it was also cheaper than the other courses.

These examples show how financial considerations can also influence the final decision of which subject to study and that entrepreneurship is cheaper than many other courses.

Employment since graduation

At the time of being interviewed, 62% of the graduates were in paid employment, of whom over half were also running their own business (see Table 17.2). As a further 29% were only engaged in running a business, this means that two-thirds of the graduates were entrepreneurs. Only 9% reported that they were currently unemployed. Those who were employed were in a range of occupations, including working in banks, for NGOs, and in schools. While it might seem surprising that so many graduates of entrepreneurship are in paid employment, respondents indicated in many cases that they saw employment as a stepping stone to setting up their own business. For some it was a matter of saving enough capital to set up a business, while for others it was also a case of developing networks and seeking inspiration. As Tom indicated, "The mentality is that you first work maybe like for two or three years, get money and then roll out to self-employment." Richard (aged 24), who was employed by an NGO that supplies communities with insecticides and mosquito nets, is an example of a graduate who had yet to set up a business. He explained how

> I hope to have my own business, manage it and employ others. I have not yet started my own business because of the lack of start-up capital. My future plans include acquiring land, building houses and shops for renting, and owning my own business.

Graduates who were self-employed had established a range of businesses, including making jewellery, event management, hair salons, dealing in wholesale fruit and vegetables, refining honey, moneylending, and real estate. Robert (aged 28) had started an events business together with his brother within three

Table 17.2 Employment status of entrepreneurship graduates (%)

Status	Number of students
Employed by others	25
Employed by others and self-employed	37
Self-employed	29
Unemployed	9
Sample size	*350*

Source: Questionnaire survey (2013).

months of graduating. He detailed how "We sat one morning, planned together and raised money to start with a tune of 10 million shillings, identified a place, bought a few things including computers and cameras, and invested the rest in marketing." Many of the businesses, however, that these young graduates establish are on a much smaller scale. Jane, for example, explained how she started off her poultry business with just three hens because she felt a need to set up a business, having studied entrepreneurship:

> When I went home [from my job] it kept disturbing me. I really felt I needed to do something. I was not doing justice to the course I did and every time you go for interviews they would ask, "You did entrepreneurship but what have you done? You are an entrepreneur do you have any business?" And you would say no, so I just went home got some money and started with three hens. I bought each at 5,000 shillings so it cost like 15,000 shillings to buy the three hens, I constructed a temporary shelter for them and that is how I started the business.

By the time of the interview, Jane had increased the number of hens in her business to around 20. Her case also illustrates how, like many other graduates, she combined being employed with running her own business.

Those graduates who were unemployed complained of the difficulty of finding employment and highlighted that they could not raise the capital to set up their own business. As found in other contexts, unemployment is increasingly common among university graduates in the global South (Jeffrey, 2010), though it remains relatively low among entrepreneurship graduates in Uganda. Many of those who were unemployed still had high hopes of starting a business soon.

Impact of entrepreneurship education on entrepreneurial intentions

We now turn to consider how graduates' entrepreneurship education had influenced their entrepreneurial intentions. Building on Ajzen (1991), we examine in turn the impact of attitudes, subjective norms, and perceived behaviour control on the graduates' decision making.

Attitudes towards entrepreneurship

The qualitative interviews revealed how studying entrepreneurship at university had changed the attitudes of almost all the graduates towards entrepreneurship. For Jacob (aged 33), studying entrepreneurship at university directly affected his entrepreneurial intentions, as he indicates here:

> before the study programme I did not know much and I expected to come and do my course and get employed in a management department or company but … by the time I finished the course I was turning to self-employment. So the

course changed my thinking. Actually I had my older brother I imitated as a role model and he was employed by an NGO, so I wanted to be like him but by the time I finished the course I had changed my mind. He even offered me a very good job in TASO [The AIDS Support Organisation] but I could not go for it.

For other graduates, although they had always intended to run a business on graduating, the course had increased their awareness of entrepreneurial opportunities. As Juliet explained,

Before starting on the course ... my mind used to be stuck on a few things but of late all the time I am moving. When I see an empty space I imagine: Can someone put a washing bay here? Who is the owner of this land? All the time my brain is thinking of more business.

Many graduates indicated that it was the knowledge they had gained from the programme that had increased their desire and ability to start up a business. They mentioned having gained a wide range of skills from the course, including being innovative, being proactive and taking up an opportunity immediately, learning how to do a feasibility study and write a business plan, record keeping, good time management, team building and networking skills, and that they should aim to start small and grow in the future. Raymond (aged 27) told how,

After the programme in entrepreneurship I made a business plan and knew how to implement it ... I had learnt how to budget and use the budget in the business plan. I did a feasibility study for a hair salon ... and the entrepreneurship education was helpful especially in managing finances ... I was able to start a hair salon in a place where there are no other big salons to outcompete mine. My friend and I collected money and bought a full hair salon on goodwill. We found an opportunity when we were looking for the place to start another business. I have now improved, added on skills, become more creative [and am] still moving and thinking of more business.

The expression "moving", used by both Juliet and Raymond, is commonly used in Uganda to refer to thinking of new ideas and getting on in life, and reflects how an entrepreneurship education can facilitate this. This is in line with many reports of how education can change young people's attitudes, skills, and lives positively (Elliott, 2011; Pellegrini, 2005).

Subjective norms regarding entrepreneurship

As outlined above, many of the graduates had been influenced by others – including peers, family members, and teachers – to study entrepreneurship in the first place. Subsequently, they all experienced views regarding their choice of degree, both during their studies and after graduating, many of which were negative. John (aged 30) indicated how his peers who were studying subjects such as law and engineering regarded studying entrepreneurship a sign of being a failure. Similarly, Samuel (aged 29) revealed how

> The programme being new, it raised a lot of issues with employment especially the career: Where will you work? What are you after this programme?... Is industry aware of the programme? So ... it sounded like negative by the statements they used to make. Among my peers the nomenclature of the programme was an issue. It had an element of small business management so they would say, "You are studying to manage kiosks and small shops." That was the attitude.

Entrepreneurship graduates explained how they and many of their classmates objected to being referred to as "people studying small business". Over time, however, many of the entrepreneurship students found that the views of others bothered them less. As Mathew (aged 26) explained,

> At first most people thought that entrepreneurship was not a marketable programme. Other people said, "You are wasting your money, where are you going to apply it?" They would give an example of government, "Where are you going to work? You studied entrepreneurship and small business management, where is it applicable?" And in the first year of study we were discouraged somehow but eventually as we collected more knowledge on what we were studying we developed more interest in doing things on our own and we were encouraged.

Part of the reason for the negative views is because degrees in entrepreneurship are relatively new in Uganda and, as Byabashaija and Katono (2011) found, university students are not expected to go into business on graduating. Over time, as graduates are able to show that they have benefited from their education, the perception of others should improve. In some instances, however, graduates found they continued to experience negative reactions even after setting up a business. Mathew explained how others look down on him because his business involves selling pineapple and sugarcane in the market, which people without a university education also do. Moses faced similar prejudice from family members:

> When you talk to any parent like to my daddy, when I told him that I am working on a project to do with cook stoves he found it very weird. He is like, "You go to university and you do cook stoves?" Yet it is making a lot of money and it is really good.

Juliet had a similar experience initially in relation to her moneylending business:

> At first when I started business ... people, plus my friends where I was sitting, used to discourage me saying, "A lady like you, how can you sit in a

kiosk?" They were so discouraging.... From there when I started a big shop and I started employing others I felt I had done it. At last I was my own boss and a boss to others.

Once her business had grown, Juliet found that other people's attitudes towards her changed for the better, showing that subjective norms regarding entrepreneurship can change over time. Despite the generally negative comments of others within their social networks, the young people had persevered with their studies of entrepreneurship and many had continued with and/or set up new businesses. While it is not possible to know how many young people are put off by other people's views from studying entrepreneurship in the first instance, as Krueger, Reilly, and Carsrud (2000) also find, subjective norms do not necessarily influence entrepreneurial intentions significantly.

Perceived behavioural control in relation to entrepreneurship

Most entrepreneurship graduates believed they were capable of starting and running their own business, leaving only a few who still felt they did not have the necessary skills. While some had always believed in their ability to run a business – even before university – many graduates emphasised how the skills they had learnt during their entrepreneurship course had greatly enhanced their self-confidence as entrepreneurs. As Mathew explained,

> To me as an individual I think the course helped me to build a lot of self-confidence. I am able to take up something even when others fear to do it. I only need to look at outcomes then I go for it, which I think other people do not always do. I have a lot of confidence in doing any business so long as it is legally acceptable.

As well as gaining skills and confidence, studying entrepreneurship also gave graduates the ability to see their way around obstacles. Although it is common for young people to cite lack of capital as a key constraint in relation to setting up a business, several of the graduates indicated how they had been taught that "When you are beginning a business, capital is not a problem, the problem is an idea." Hence, they no longer saw capital as a constraint. Mathew, for example, outlined how he believed young people could always get money one way or another to set up a business:

> There are other ways of accessing money other than banks. You can contact friends – they are a source of capital – or relatives. You can form into groups and make contributions and form companies ... so there are very many ways of getting money. Other people if they can get opportunities to be employed they can work for a short period and get some little money to start business.

Furthermore, as John, who runs a real estate business, indicated, he had learned through his studies that "You can control your own destiny." Not all the graduates, however, demonstrated such a high level of perceived behavioural control. As one of the few graduates who had never run a business, Samuel, explained:

> I have seen opportunities which I always tell people that are usually in business and they grab them. They are usually the first ones to catch so I know how to spot but seizure is the action that is still a problem.

Thus, although Samuel had improved his entrepreneurial skills through his studies, he still lacked the confidence and go-ahead to actually set up a business himself. This was not the case, though, for most graduates, for whom their entrepreneurship education had increased their belief in their ability to set up a business (see also Galloway & Brown, 2002; Garavan & O'Cinneide, 1994; Kolvereid & Moen, 1997).

Overall, this discussion of the impact of entrepreneurship education on graduates' entrepreneurial intentions indicates how their attitudes towards entrepreneurship had changed as a result of their studies. Despite subjective norms driving others to criticise their choice, they felt better equipped to take on the role of an entrepreneur.

Conclusions

As this chapter has shown, entrepreneurship education at university level in Uganda increases young people's faith in their ability to become entrepreneurs and provides them with the necessary skills to set up and run a business. Influential aspects taught on the programme include: being innovative, spotting opportunities, learning how to set up and run a business, team building, and networking skills. For many entrepreneurship graduates, however, their entrepreneurship education had started long before their degree – either while growing up in an entrepreneurial environment as their parents had run businesses or through studying entrepreneurship at school. Some had even started businesses before entering university and many more were running a business while studying. In both instances, the profits helped fund their studies.

Although it might seem surprising that most young people who have studied entrepreneurship enter paid employment on graduating, for many this is a deliberate strategy to enable them to build up capital to invest in a business and establish important networks. Many young people operate a business alongside being employed. It is thus important to recognise the ways in which young people combine education, employment, and running a business – either simultaneously or concurrently – and how these can be mutually reinforcing. For almost one-third of the graduates, being an entrepreneur is their only occupation; when combined with those who run a business alongside employment, almost two-thirds of the sample report running a business. A degree in entrepreneurship clearly enhances positive attitudes towards business and generates an entrepreneurial spirit among graduates, many of whom develop a passion for business (Frese &

Gielnik, 2014). Consequently, unemployment levels are relatively low for entrepreneurship graduates.

Despite entrepreneurship graduates valuing their university education, it is
not yet broadly accepted as a useful degree by Ugandan society. It is important
that the value of an entrepreneurship education becomes more widely recognised
so that prospective students do not face the stigma of having chosen to study
entrepreneurship. Following the evident success of the current entrepreneurship
programmes, it can be argued that entrepreneurship degrees should be offered by
more universities and that all university students should be exposed to some
level of entrepreneurship education (Byabashaija & Katono, 2011). The curric-
ulum, however, requires some restructuring: some of the young graduates complained that the degree courses were too theoretical and argued that they would
have preferred a more practical education that included more contact with successful entrepreneurs. One way of increasing the standing of entrepreneurship
education in society at large and simultaneously improving it would be for successful entrepreneurs among the current cohort of graduates to return to their
former universities to inspire undergraduates. Setting up internships in successful businesses and establishing university-led incubation centres are also ideas
expressed by the graduates that could be explored as ways of improving entrepreneurship education and generating more young entrepreneurs.

Note

1 Success cards are greeting cards with good-luck messages wishing people success in
   events like examinations.

References

Ajzen, I. (1991). The theory of planned behaviour. *Organizational Behaviour and Human
   Decision Processes, 50*, 179–211.
Alberti, F., Sciascia, S., & Poli, A. (2004, July). *Entrepreneurship education: Notes on an
   ongoing debate.* Paper presented at the 14th Annual International Entrepreneur Conference, University of Napoli Federico II, Naples.
Byabashaija, W., & Katono, I. (2011). The impact of college entrepreneurial education on
   entrepreneurial attitudes and intention to start a business in Uganda. *Journal of Devel-
   opmental Entrepreneurship, 16*(1), 127–144.
DeJaeghere, J., & Baxter, A. (2014). Entrepreneurship education for youth in sub-Saharan
   Africa: A capabilities approach as an alternative framework to neoliberalism's individ-
   ualizing risks. *Progress in Development Studies, 14*(1), 61–76.
Elliott, J. (Ed.). (2011). *Reconstructing teacher education.* London: Routledge.
Frese, M., & Gielnik, M. M. (2014). The psychology of entrepreneurship. *Annual Review
   of Organizational Psychology and Organizational Behaviour, 1*, 413–438.
Galloway, L., & Brown, W. (2002). Entrepreneurship education at university: A driver in
   the creation of high growth firms? *Education + Training, 44*, 398–404.
Garavan, T. N., & O'Cinneide, B. (1994). Entrepreneurship education and training programmes: A review and evaluation – Part 2. *Journal of European Industrial Training,
   18*(11), 13–21.

Gnoth, J. (2006, September). *Perceived motivations and barriers for entrepreneurship amongst NZ students*. Paper presented at the Rencontres de St-Gallon Understanding the Regulatory Climate for Entrepreneurship and SMEs, Wildhaus.

Gorman, G., Hanlon, D., & King, W. (1997). Some research perspectives on entrepreneurship education, enterprise education and education for small business management: A ten-year literature review. *International Small Business Journal, 15*(3), 56–77.

Honig, B. (2004). Entrepreneurship education: Toward a model of contingency-based business planning. *Academy of Management Learning and Education, 3*(3), 258–273.

International Labour Organization. (2009). *Supporting entrepreneurship education: A report on the global outreach of the ILO's Know about Business Programme*. Geneva: International Labour Office.

Jeffrey, C. (2010). *Timepass: Youth, class, and the politics of waiting in India*. Stanford, CA: Stanford University Press.

Kagoda, A. M. (2011, October). *Assessing the effectiveness of affirmative action on women's leadership and participation in education sector in Uganda*. Paper presented at the IIEP Policy Forum on Gender Equality in Education, Paris.

Kolvereid, L. (1996). Organizational employment versus self-employment: Reasons for career choice intentions. *Entrepreneurship Theory and Practice, 20*(3), 23–31.

Kolvereid, L., & Moen, Ø. (1997). Entrepreneurship among business graduates: Does a major in entrepreneurship make a difference? *Journal of European Industrial Training, 21*(4), 154–160.

Krueger, N. F. (1993). The impact of prior entrepreneurial exposure on perceptions of new venture feasibility and desirability. *Entrepreneurship Theory and Practice, 18*(1), 5–21.

Krueger, N. F., Reilly, M. D., & Carsrud, A. L. (2000). Competing models of entrepreneurial intentions. *Journal of Business Venturing, 15*(5–6), 411–432.

Kuratko, D. F. (2003). *Entrepreneurship education: Emerging trends and challenges for the 21st century* (White Paper Series). Washington, DC: Coleman Foundation and US Association of Small Business and Entrepreneurship.

Morales-Gualdrón, S. T., Gutiérrez-Garcia, A., & Dobón, S. R. (2009). The entrepreneurial motivation in academia: A multidimensional construct. *International Entrepreneurship and Management Journal, 5*(3), 301–317.

Pellegrini, A. D. (2005). *Recess: Its role in education and development*. London: CRC Press.

Ronstadt, R. (1987). The educated entrepreneurs: A new era of entrepreneurial education is beginning. *American Journal of Small Business, 11*(4), 37–53.

Sandhu, M. S., Sidique, S. F., & Riaz, S. (2011). Entrepreneurship barriers and entrepreneurial inclination among Malaysian postgraduate students. *International Journal of Entrepreneurial Behaviour and Research, 17*(4), 428–449.

Solomon, G. T., Duffy, S., & Tarabishy, A. (2002). The state of entrepreneurship education in the United States: A nationwide survey and analysis. *International Journal of Entrepreneurship Education, 1*(1), 65–86.

Thurik, R., & Wennekers, S. (2004). Entrepreneurship, small business and economic growth. *Journal of Small Business and Enterprise Development, 11*(1), 140–149.

Uganda Bureau of Statistics (UBOS). (2010). *Statistical abstract*. Government of Uganda.

UNESCO Institute for Statistics. (2011). *Global education digest 2011: Regional profile – Sub-Saharan Africa*. Retrieved 5 June 2012 from www.uis.unesco.org/Education/Documents/GED2011_SSA_RP_EN.pdf.

Wang, C. K., & Wong, P.-K. (2004). Entrepreneurial interest of university students in Singapore. *Technovation, 24*(2), 161–172.

Wilson, F., Kickul, J., & Marlino, D. (2007). Gender, entrepreneurial self-efficacy, and entrepreneurial career intentions: Implications for entrepreneurship education. *Entrepreneurship Theory and Practice, 31*(3), 387–406.

Concluding comments to Part V

Katherine V. Gough and Thilde Langevang

The preceding three chapters examined different support structures and schemes initiated to stimulate youth entrepreneurship. While varying in focus and scope, the chapters all showed the importance of understanding the particular situation of specific groups of young people and the need to be innovative when developing holistic, flexible, and tailor-made support mechanisms to suit their needs. Despite highlighting the benefits stemming from the respective support schemes and structures, the chapters also indicated some key limitations that need to be considered.

The importance of paying attention to the social context that young people and their access to support are embedded in was highlighted in a Zambian context. In a situation with limited formal support structures, young entrepreneurs rely extensively on the resources they can mobilise through their own social networks. By drawing on their social capital, young entrepreneurs operating in market places discover new business opportunities and gain access to information, finance, skills, and customers. Formal institutional support to young entrepreneurs, however, can be successful. Microfinance institutions (MFIs) in Kampala have designed their services and products to match the particular needs of young people. By adopting a holistic approach that did not respond to young people's financial needs in isolation from their other needs, the MFIs were able to offer a variety of services and products to build the capacities of the youth.

The importance of understanding the particular life situation of young people in specific settings was also revealed in the discussion of entrepreneurship education. While entrepreneurship training at educational institutions is typically seen as preceding business start-up (see, for example, Schoof, 2006), some university students of entrepreneurship in Kampala actually start operating businesses before they begin their studies, and others start their businesses during their studies. This is important knowledge for educational institutions to consider as many students consequently arrive at university with prior business experience. Innovative ways of capitalising on their experiences in the classroom, such as using their existing businesses as practical live cases, would be beneficial for all students. The need for being innovative when developing support schemes for young people also appeared clearly in the analysis of the Ugandan MFIs and showcased the importance of a supportive entrepreneurial environment within an organisation for successful implementation of the innovations.

Turning to consider some limitations of the respective support mechanisms, while young people do indeed depend on social networks in their business operations and are able to gain access to key resources via these, young entrepreneurs are restricted in what they can accomplish through drawing solely on their social capital. There are also downsides to relying heavily on social capital as it can have detrimental effects on young people's businesses in situations where they spend time and energy on developing relationships that turn out to be unrewarding. These findings tie in with other critical studies, showing that the type of social capital that young entrepreneurs in the global South can use may enable them to "get by" but not "get ahead" (Turner & Nguyen, 2005). A key limitation is that young people rely predominantly on strong ties to a close network and only to a limited extent on weaker ties that could link them to actors with more resources as well as to formal institutions. Importantly, the existence of social capital in low-income communities should not be an excuse for the government and other stakeholders not to provide formal support structures and services for young entrepreneurs.

The microfinance case from Kampala demonstrated that it is actually possible for MFIs to reach the youth and revealed how the young clients generally have positive experiences of the services and products provided. The study illustrated, however, that the issue of sustainability of the programmes as well as of the livelihood activities of the youth is a key issue MFIs face. Time will show if the institutions are able to sustain their programmes but the study indicated that some young people had difficulties making the deposits required, while others found the loans to be too small. In order to develop effective microfinance schemes, the definition of "sustainability" needs to be broadened from a narrow focus on programme sustainability to generating viable livelihoods through enterprise promotion (Chigunta, 2001). Furthermore, since microfinance is only relevant for entrepreneurs running very small enterprises, there is a need to develop innovative products and services to cater for young entrepreneurs endeavouring to expand beyond the micro stage.

Turning to entrepreneurship education at university level, although it provides graduates with the skills and confidence needed to operate their own businesses, not all students feel well equipped to start a business after their studies and some voiced critique that the courses were too theoretical with limited practical content. This highlights the need to make sure that the focus of entrepreneurship education is on teaching "for entrepreneurship" rather than "about entrepreneurship" (Mwasalwiba, 2010). As this book has shown, the key issue in a sub-Saharan African context is not about creating awareness of the benefits of entrepreneurship among young people but rather providing young entrepreneurs with the skills and wherewithal to operate effective businesses. Doing so would not only help facilitate young people's transitions into adulthood but also contribute to economic growth and poverty reduction.

References

Chigunta, F. (2001). *Youth livelihoods and enterprise activities in Zambia*. Report to IDRC, Canada.

Langevang, T., Namatovu, R., & Dawa, S. (2012). Beyond necessity and opportunity entrepreneurship: Motivations and aspirations of young entrepreneurs in Uganda. *International Development Planning Review, 34*(4), 439–460.

Mwasalwiba, E. S. (2010). Entrepreneurship education: A review of its objectives, teaching methods, and impact indicators. *Education+Training, 52*(1), 20–47.

Schoof, U. (2006). *Stimulating youth entrepreneurship: Barriers and incentives to enterprise start-ups by young people* (No. 388157). Geneva International Labour Organization.

Turner, S., & Nguyen, P. A. (2005). Young entrepreneurs, social capital and Doi Moi in Hanoi, Vietnam. *Urban Studies, 42*(10), 1693–1710.

Index

Page numbers in *italics* denote tables, those in **bold** denote figures.

capital: accessing 90, 102 (*see also* access to finance); sources of 71, 102, 141–2, 153–5; use of the body as 88

cash crops: eastern Uganda 135; northern Ghana 121, 122; northern Zambia 152; production of as example of entrepreneurial activity 134; risk to food security 137; southwestern Uganda 195

challenges: access to finance 223–4, 238; inadequacy of education 89; key challenges affecting youth entrepreneurship in Nima-Maamobi *101*; and opportunities for young entrepreneurs in Uganda 141–3; poor infrastructure 89; regulations 89; for young entrepreneurs in Accra 101–5; for young entrepreneurs in Lusaka 72–4

characteristics of youth entrepreneurship: Ghana 34–40; Uganda 20–7; Zambia 50–4

charcoal production/selling 74, 118, 138, 150, **151**, 152–3

Chawama, Zambia 67–79; education levels 73; employment status of young people in *70*; entrepreneurial activities of young people 70–2, 74–5; knowledge and use of government programmes 73; overview 68; renting out gumboots **75**; skills training centre 76–7

chilimba (informal revolving fund) 74, 232

chitemene (shifting cultivation system) 146, 148, 150

choice of business, reasons for *141*

climate change, impact on the handicraft sector 195

comparative approach, centrality to the YEMP research project 7

competition: strategies for handling 72; young entrepreneurs' concerns 90

cooperatives 141, 152, 153, 154, 227, 238

corporate social responsibility 77

corruption 42, 60, 73, 89

credit: access to in the informal sector 230; social capital and 227

disabilities, and the handicraft sector 197

distant opportunity spaces 119, 121, 125, 129

diversification 88, 92, 154, 182, 196, 203, 214

economic growth: Ghana 32, 94; role of entrepreneurship in 15

education: and entrepreneurial performance 38; Ghanaian policy

outlook 41; inadequacy of as challenge to entrepreneurs 89; role of in the development process 248

education for entrepreneurship *see* entrepreneurship education

education for women, and social change 36

education levels: gendered perspective 170; Ghana 38, 170; handicraft sector 197; mobile telephony sector 170; small-scale mining 184, 187; status of in the global South 248; Uganda 20, 24, 89, 94, *98*, 99, 197; Zambia 52, 72–3

employment: preferences of young people 72; provided by young entrepreneurs *25*, *40*, 54, *89*, 127, 213–14; status of young people *70*, 83, *84*, 99

entrepreneurial activities: agricultural production as enabler of 22; examples of 1–11, 18–31, 32–47, 48–58, 67–79, 80–93, 94–107, 117–31, 132–45, 146–58, 167–80, 181–92, 193–205, 206–16, 225–36, 237–47, 248–62; gendered perspective 118; *see also* illegal activities; nonfarm activities

entrepreneurs: heroic depiction 223; typology 16

entrepreneurship: attitudes towards *23*, 37–8; the concept 3, 68; as an "everyday societal phenomenon" 4; global Northern focus of research 3; a key argument for promoting 39; perceptions of 70; process-oriented character 4; promotion of as driver of social/economic transformation 1; public perceptions 72; research in the global South 4; role of in economic growth 15; topics of research 4

entrepreneurship education: and attitudes towards entrepreneurship 248–62; defining 249; employment status of graduates 254–5; entrepreneurship teaching **250**; existing research 248, 249–50; gendered perspective 251; government promotion 249; impact on entrepreneurial intentions 255; increased focus on 224; motivations for studying 252–4; overview 249–50; and perceived behavioural control 258–9; profile of graduates 251–2; and subjective norms 256–8; universities offering 250

entrepreneurship rates (TEA) in Ghana, Uganda, and Zambia compared to regional averages **60**

illegal activities: smuggling in northern
Ghana 125; undertaken by youth in
Nima-Maamobi 99; young people's
justifications for 125
images of young entrepreneurs:
entrepreneurship teaching **250**; farming
sector **137, 151**; handicraft sector **197,
199**; hotel and lodgings sector **212**;
mining sector **184, 185**; mobile
telephony sector **176, 178**; retail sector
84, **123, 139, 156**; scrap collector 90;
transporting charcoal **151**; young men
returning from Ivory Coast **126**
informal sector: and access to credit 230;
business types 194; the concept 4;
gendered perspective 169, 218;
Ghanaian perspective 37; as main source
of employment in Africa 65; Ugandan
perspective 19; Zambian perspective 50,
226
information asymmetries 238, 246
infrastructure 81, 92, 101, 103, 110, 152
innovation: examples 74–5; MFIs and the
need for youth tailored innovations
238–40 (*see also* MFI innovations);
research on 4
international trading, by young
entrepreneurs in Uganda 26
Ivory Coast, goods smuggled into Ghana
from 125

job growth, aspirations for *26, 40*
job-creation, Uganda 25

kaloba (loans from illegal moneylenders)
232
Kampala 80–95; youth entrepreneurship
in: challenges 88–91; diversification 88;
employees *89*; employment status of
young people 83, *84*; entrepreneurial
activities 81, 84; expansion aspirations
91; and family obligations 86; growth
inhibiting factors 89; images of young
entrepreneurs **83, 84, 90**; motivating
factors 86; paths to enterprise ownership
83–6; resource management 87–8;
sources of institutional support 87; study
location 80–1 (*see also* Bwaise); study
methodology 81–3
kayayoo (female porters) 127
Kenya, growth in the handicraft sector 194

land, access to *see* access to land
land tenure, categories of 147

livelihood strategies, of young Ghanaians:
context of study 117–19; farming and
fishing 121–3; gendered perspective
118; illegal activities 125; migration
125–9; nonfarm activities 118, 123–5;
shops in Kiape 125; shops in Mankuma
123, 125; study location 118, 119, **120**,
121; study methodology 119–21; of
young people in rural Zambia 46
loans 71, 73–4, 87, 92, 109, 142, 155, 214,
232, 243
Lusaka: living conditions 67; overview 68;
urban population 68
Lusaka 67–79, 225–36; youth
entrepreneurship in, awareness and
perceptions of enterprise development
programmes 73; and belief in witchcraft
74; business association membership 74;
challenges for 72–4; Chawama Youth
Project 76–7; employment status of
young people in Chawama *70*;
entrepreneurial activities 70–2, 74;
gendered perspective 70; and
governmental regulation 73;
inventiveness 74–5; sector distribution
71; study aim 69; study location 68;
study methodology 69

Malawi 15, 68, 93
manufacturing 25, 38–9, 53, 71, 109, 133,
135–6, 138
marginalisation, of young people and its
impact 1
mentorship 74
"merry-go-round" schemes 142
methodologies of studies 1–11, 18–31,
32–47, 48–58, 67–79, 80–93, 94–107,
117–31, 132–45, 146–58, 167–80,
181–92, 193–205, 206–16, 225–36,
237–47, 248–62; Accra study 98;
entrepreneurship education study 250–1;
female entrepreneurship study 195–6;
Kampala study 81–3; livelihood
strategies study 119–21; Lusaka study
69; MFI innovations study 240; mobile
telephony sector study 169; rural
entrepreneurship studies 135–6, 148–9;
small-scale mining study 182–3; social
capital study 227–9; tourism industry
study 207–8
MFI innovations: case studies 240–3;
holistic approach to serving youth 241–2;
providing decent housing for youth
242–3; youth savings initiative 241

resource management, examples of 87–8
retail activities, examples of 74
revolving funds 187, 200–1, 232
risk-spreading strategies: mining sector
190; nonfarm sector 140, 146, 152;
small-scale mining sector 190; *see also*
diversification
rotational savings schemes 142, 201–2
rural areas, in-migration into 157
rural entrepreneurship, Uganda 132–45;
challenges and opportunities 141–3;
combining activities 140; farming
136–8; migration 140; nonfarm
activities 138; rural livelihoods 132–5;
study location **134**, 135; study
methodology 135–6, 148–9
rural entrepreneurship, Zambia 132–45;
combining farming and small-scale
business 154–5; demographic
perspective 149–52; farming 150,
153–4; household composition *149*;
livelihood strategies 153–7; migration
155–7; nonfarm activities 150; rural
livelihoods 148; sources of household
income *150*; study location **148**; study
methodology 148–9
rural entrepreneurship, Ghana 117–31
rural locations, advantages and
disadvantages for entrepreneurial
activity 142–3
rural youth: access to land 137;
cooperative membership 152, 153, 154;
livelihood activities 121–9; livelihood
strategy comparisons 159–61; options
for 136

savings, as source of start-up capital 189
sector classifications 38
sector distribution of TEA entrepreneurs:
Ghana *25*, *39*; Uganda *25*; Zambia 52,
53
services provided by young entrepreneurs
71
shifting cultivation 121, 146–7
shop owning/trading, as livelihood activity
154
Skilling Uganda Programme 27
skills training: Chawama Youth Project
provision 76–7; Ghanaian policy
outlook 41; importance of 86; received
by young entrepreneurs in Kampala 87;
Ugandan initiatives 27–9; Zambia 55–6
Slum Cinema 82
small shops, as livelihood activity 154–5

small-scale farming: farmer in his garden
137, **151**; government intervention in
northern Zambia 146–7
small-scale mining (SSM) 181–92;
attractions of 119, 122, 165, 190;
benefits of a supported SSM sector 191;
benefits of for northern Ghana 127;
buyers' role and requirements 187;
carrying ore to processing sites **185**;
catering services 188; chiselling ore
184; deaths 185; definition 181–2; and
education levels 184, 187;
entrepreneurial activities in mining
183–7; entrepreneurial activities
supporting mining 188, 190; estimated
participants 182; the extraction process
183; gendered perspectives 184, 186–9;
hazards 185, 187; international
recognition of the importance for youth
190; legal perspectives 181, 189; as
means of raising start-up capital 185–6;
media campaign against 181; ore
processing 186; percentage of gold
production accounted for in Ghana 182;
pit owners 186; problems associated
with 128; risk-spreading strategies 190;
robbery 190; scavenging waste ore 186;
as source of start-up capital 188; sources
of start-up capital 189; sponsors' role
and requirements 187; Tanzania study
187; training provision 183;
transportation of miners and gold 189;
as viable livelihood option 181; women
carrying ore to processing sites **185**
social capital: and access to finance 230–2;
and access to resources 223; basic
assumption 225; Bourdieu's definition
225–6; and business information
acquisition 229–30; the concept of 201,
225; customer relations 234–5; defining
225–6; examples of strong ties 229;
gendered perspective 189; handicraft
sector 201–2; skills acquisition 233 (*see
also* skills training); in Zambia 225–7
social embeddedness, Granovetter's notion
226, 235
social entrepreneurship 67–8, 76–7, 110
social intermediation: crux of 244;
function of MFIs 238
social mobility, importance of
entrepreneurship for 99
social networks 4, 40, 87, 90–1, 93, 95,
136, 201, 215, 225, 258
South Africa 15

start-up capital: as constraint to youth
entrepreneurship 102–3; sources of
53–4, 71, 141–2, 153–5, 185–6, 188–9,
211, 215; *see also* access to finance
structural adjustment policies (SAPs),
impact of implementation in urban areas
65
sub-Saharan Africa: rural population 115;
urban growth and its impact 65
sugarcane farming, Uganda 135–6, 138

Tanzania, small-scale mining study 187
taxation 4, 89, 104, 182
TEA (total entrepreneurial activity):
function 21; Ghana 39; as key
measurement 16; Uganda 21; Zambia 52
TEA entrepreneurs: by age-group *36*; five-
year job growth aspirations *26, 40*; by
gender **35**, 36; regional perspective
22–3; sector distributions *25, 39*
tourism industry 206–16; employment
conditions 206–16; employment
opportunities 208–14; gendered
perspective 212, 214; growth of in sub-
Saharan Africa 166; hospitality sector,
advantages of employment 209; hotels
and lodging subsector 209–11; impact
on the handicraft sector 194–6;
network building opportunities 209;
potential for youth employment 209;
as source of capital 215; supply
subsector 211–14; training
opportunities 209–10; in Uganda 207;
young man supplying drinks to a hotel
212
traders' associations, membership of 74,
155
trading: as livelihood strategy 155–6; in
mining settlements 188; in northern
Zambia 150, 155–6; reason for
popularity of 100
trading activities: young entrepreneurs in
Accra 100; young entrepreneurs in
eastern Uganda 138; young
entrepreneurs in Kampala 84; young
entrepreneurs in Lusaka 71
training opportunities: mobile telephony
sector 174–5; small-scale mining sector
183; tourism industry 209–10; *see also*
skills training
transport 147–8, 152, 188

Uganda 18–31, 67–79, 117–31, 193–205,
206–16, 237–47, 248–62
Uganda Youth Welfare Services (UYWS)
81–2
unemployment: benefits of a supported
SSM sector for reducing 191;
measurement challenges 4;
unemployment rates, young people 1
urban migration, impact on the handicraft
sector 195
urbanisation, in sub-Saharan Africa 65

vegetable production, as source of start-up
capital 153
vocational education 27–8, 34
vulnerability, social capital as strategy
against 226

wage employment, and entrepreneurship
24, 39, 54
witchcraft 74, 91
women: access to land 39; attitudes
towards entrepreneurship 23; education
for and social change 36; empowerment
of by MFIs in Uganda 239; expectations
of in rural Africa 193; gaining access to
capital 204; key aspect of social capital
201; levels of established business
ownership 26; as primary focus for
MFIs 237, 238
women entrepreneurs, and social change
36
World Bank 5, 44, 49, 167, 190

youth: defining 2, 69, 104; focus of
cultural approaches 2; life-stage/
transitions approach vs the cultural
approach 2–3; sub-Saharan
categorisation 2; UN categorisation 2
Youth Entrepreneurship Facility, Uganda
28
youth entrepreneurship schemes,
politicisation of in Africa 60
"youth in peril" 115
Youth Livelihood Programme, Uganda 28
youth transition studies, criticisms 2
Youth Venture Capital Fund, Uganda 28

Zambia 48–58, 67–79, 146–58, 225–36
"zigzagging" 81, 85, 92, 110
Zimbabwe, tourism study 209

Printed in the United States
By Bookmasters